Lecture Notes in Chemistry

Edited by G. Berthier, M. J. S. Dewar, H. Fischer
K. Fukui, G. G. Hall, H. Hartmann, H. H. Jaffé, J. Jortner
W. Kutzelnigg, K. Ruedenberg, E. Scrocco, W. Zeil

20

Benjamin Fain

Theory of Rate Processes in Condensed Media

Springer-Verlag
Berlin Heidelberg New York 1980

Author

Benjamin Fain
Institute of Chemistry
Tel-Aviv University
61390 Ramat-Aviv
Tel-Aviv/Israel

ISBN-13: 978-3-540-10249-6 e-ISBN-13: 978-3-642-93153-6
DOI: 10.1007/978-3-642-93153-6

Softcover reprint of the hardcover 1st edition 1980

PREFACE

A wide range of important physical and chemical phenomena may be named by the same title: Rate processes in condensed media. They have the same underlying physics and mathematics. To these phenomena belong:

1) Small polaron motion [a]
2) Electron transfer between ions in solutions [b] and in photosynthetic centers [c].
3) Electronic energy transfer between molecules or ions in solids and in liquids [d].
4) Enzymatic catalysis [e]
5) Group transfer in biological systems [f].
6) Electron-hole recombination in semiconductors [g].
7) Non-radiative electronic relaxation in ionic centers [h] and in impurity states in insulators [i].
8) Recombination in amorphous solids [j].
9) Radiationless transitions in large molecules [k].

At present a unified theoretical and conceptual framework exists for description and understanding of all these diverse phenomena.

Our aim in writing this work is to introduce the student and the research worker in chemical physics to the main aspects of the conceptual framework of rate processes in condensed media. There exists an extensive literature devoted to various rate processes in condensed media. Among recent works, the books of Levine and Jortner [l], Fong [m] and Ulstrup [n] should be mentioned. The existence of this literature enables us to concentrate on the major theoretical aspects, omitting description and presentation of experimental data.

We tried to make our work self-contained. Almost all the information necessary for reading the book is included. References used in the work by no means cover the literature on the subject.

The first chapter provides the quantum mechanical basis necessary for the next chapters. General concepts of quantum theory and some specific applications are given. They include the introduction of the density matrix, descriptions of two-state systems, probability per unit time and the Born-Oppenheimer approximation.

Chapter II is devoted to the basic quantum mechanical models used in the theory of rate processes in condensed media.

Chapter III is devoted to the derivation from the first principles of the equations describing rate processes in condensed media, in particular, the range of the applicability of the master equations is clarified.

In Chapters IV-VII particular model systems are explored.

In conclusion, I would like to mention that this work is based on a course of lectures given in the Chemistry Department of Tel-Aviv University.

I would like to thank Professor Joshua Jortner for many stimulating and helpful discussions. This research is supported by the Center for Absorption in Science, The Ministry for Immigrant Absorption, State of Israel.

I would also like to thank Dr. Ephraim Buhks for discussions and assistance in preparing the manuscript.

I would finally like to thank Mrs. Louise Fattal for editing and typing the manuscript.

---

a) T. HOLSTEIN, Ann. Phys. 8, 343 (1959).

b) R.A. MARCUS, J. Chem. Phys. 24, 966 (1956); V.G. LEVICH, Adv. in Electrochem. Electrochem. Eng. 4, 249 (1965); N.R. KESTNER, J. LOGAN and J. JORTNER, J. Phys. Chem. 78, 2148 (1974).

c) J. HOPFIELD, Proc. Nat. Acad. Sci. USA 71, 3640 (1974); J. JORTNER, J. Chem. Phys. 64, 4860 (1976).

d) R. FÖRSTER, Naturwissenschaften 33, 166 (1946); D.L. DEXTER, J. Chem. Phys. 21, 836 (1953).

e) M. DIXON and E. WEBB, Enzymes (Longman, N.Y. 1964); B. FAIN, J. Chem. Phys. 65, 1854 (1976).

f) R.H. AUSTIN, K.W. BEESON, G. EISENSTEIN, H. FRAUENEFELDER and I.C. GARSIALUS, Biochemistry N.Y. 14, 5355 (1975).

g) R. KUBO and Y. TOYOZAWA, Prog. Theor. Phys. 13, 160 (1955); C.H. HENRY and D.V. LANG, Phys. Rev. B, 15, 989 (1977).

h) R.H. BARTRAM and A.M. STONEHAM, Solid St. Comm. 17, 1593 (1975); M.D. STURGE, Phys. Rev. B, 8, 6 (1973).

i) J. JORTNER, Vacuum Ultraviolet Radiation Physics, ed. by E.F. KOCH, R. HAEMSEL and C. KUNZ (Pergamon-Vieweg) p. 263 (1976); J. Chem. Phys. 64, 4860.

j) N.F. MOTT, N.F. DAVIS and R.A. STREET, Phil. Mag. 32, 961 (1975).

k) R. ENGLMAN and J. JORTNER, Molec. Phys. 18, 145 (1970).

l) Molecular Energy Transfer, ed. by R. LEVINE and J. JORTNER, Wiley Int. New York (1976).

m) Theory of Molecular Relaxation by F.K. FONG, Wiley Int. New York (1975).

n) Charge Transfer Processes in Condensed Media, by J. ULSTRUP, Springer-Verlag Berlin (1979).

Table of Contents

I. BASIC QUANTUM-MECHANICAL EQUIPMENT

1.1 Basic concepts of quantum theory

Quantum theoretical description of quantum objects differs essentially from the classical description. Quantum theory, unlike the classical theory, is a statistical theory in principle. The laws of quantum theory do not govern the actual behavior of a particular object, but give the probability of the various ways in which the object may behave as a result of an interaction with its surroundings. Several postulates form the basis of the quantum description of physical phenomena.

1. Each physical quantity has corresponding to it a linear Hermitian operator or matrix. For example, the radius vector of a particle $\vec{r}$ is associated with multiplication operator $\vec{r}$, the momentum of the particle - with the operator $\vec{p} = i\hbar\vec{\nabla}$, the angular momentum - with the operator $\hbar\vec{L} = \vec{r}\times\vec{p} = i\hbar\vec{r}\times\vec{\nabla}$. The operators corresponding to the physical quantities are, generally speaking , not commutative. There are commutation relations between the coordinate and the momentum operators:

$$x p_x - p_x x = i\hbar; \quad y p_y - p_y y = i\hbar; \quad z p_z - p_z z = i\hbar \tag{1.1}$$

and there are also commutation relations between the operators of the components of the angular momentum

$$L_x L_y - L_y L_x = iL_z, \quad L_y L_z - L_z L_y = iL_x, \quad L_z L_x - L_x L_z = iL_y \tag{1.2}$$

where $\hbar$ is Planck's constant divided by 2π. Commutation relations such as (1.1) and (1.2) are basic characteristics of operators.

2. Only the eigenvalues of the operator A can be the result of a precise measurement of a physical quantity. The characteristic difference from classical theory is the fact that physical quantities may take up a discrete, as well as a continuous, series of values. It is well known, for example, that the energy spectrum of atoms is discrete in nature.

3. The state of the physical system can be always described by the quantity which is called the density matrix ρ. The mean value of the physical quantity represented by the operator (matrix) A is given by the formula

$$\langle A \rangle = \mathrm{Tr}(\rho A) = \sum_{nm} \rho_{nm} A_{mn} \tag{1.3}$$

where A_{mn} are the matrix elements of the operator A.

These matrix elements are calculated with the aid of some particular basis of the eigenfunctions of a certain Hermitian operator B

$$B\Psi_n = b_n \Psi_n \; ; \quad A_{mn} = \langle m | A | n \rangle = \int \Psi_m^* A \Psi_n \, dq \tag{1.4}$$

The transition from one basis of the eigenfunctions Ψ_n to another Ψ'_n of some other operator C

$$C\Psi'_n = c_n \Psi'_n \tag{1.5}$$

is the change of the representaion of the operator A:

$$\Psi'_n = \sum_m U_{mn} \Psi_m = U \Psi_n \tag{1.5}$$

Here the operator U, determining the change of the transformation, is defined by (1.5), which is just the expansion of the function Ψ'_n over the eigenfunctions Ψ_m. In the new representation (C-representation) the operator A is presented by the matrix

$$A'_{mn} = \int \Psi'^*_m A \Psi'_n \, dq = \int (U\Psi_m)^* A (U\Psi_n) \, dq =$$
$$= \sum_{kl} \int U^*_{km} \Psi^*_k A U_{ln} \Psi_n \, dq = \sum_{kl} U^*_{km} A_{kl} U_{ln} =$$
$$= \sum_{kl} (U^+)_{mk} A_{kl} U_{ln} = U^+ A U \tag{1.6}$$

where the Hermitian conjugated operator U^+ is defined by the equation

$$(U^+)_{mk} = U^*_{km} \tag{1.7}$$

From the fact that the eigenfunctions of Hermitian operators may be chosen orthogonal and normalized to unity

$$\int \Psi'^*_n \Psi'_m \, dq = \int \Psi^*_n \Psi_m \, dq = \delta_{nm} , \tag{1.8}$$

the unitary property of the operator U follows:

$$U^{+} = U^{-1} \tag{1.9}$$

The transformation properties of the density matrix may be derived from the requirement that mean values of the physical quantities should not depend on choice of the representation

$$\langle A' \rangle = \langle A \rangle = Tr(\rho' A') = Tr(\rho' U^{+} A U) = Tr(\rho A) \tag{1.10}$$

Therefore the density matrix ρ satisfies the following transformation property

$$\rho = U \rho' U^{+}, \quad \rho' = U^{+} \rho U = U^{-1} \rho U \tag{1.11}$$

The density matrix may be normalized to unity

$$\langle 1 \rangle = Tr(\rho) = \sum_n \rho_{nn} = 1 \tag{1.12}$$

It is obvious that in A-representation the matrix A_{mn} is diagonal

$$A_{mn} = A_n \delta_{mn}$$

In this representation

$$\langle A \rangle = \sum_n \rho_{nn} A_n \tag{1.13}$$

From (1.12), (1.13), and the postulate 2 claiming that the eigenvalues A_n are a result of a precise measurement of this quantity A, it follows that ρ_{nn} has the meaning of the probability to find the system in the state n.

It is worth-while to clarify at this point the meaning of the expression "the probability to find the system in the state n". Speaking about the probabilities we should define the statistical ensemble to which these probabilities refer.

First of all we should take into account the fact that the measurement or experiment, generally speaking, changes the state of the object. We cannot ignore (as it is possible to do in the classical physics) the influence of the measuring device on the object. It is therefore necessary (in order not to go outside the framework of the given ensemble) to return the object after each measurement to the original quantum state or to deal with a set of objects in one and the same quantum state. In this case the measurement is made once on each object. Then, if we have a set of objects in the same quantum state (characterized by the density matrix ρ) we may perform different measurements, corresponding to different quantities characterizing the object. Thus we may measure the coordinate of the electron, its momentum, spin, etc. Each type of the measurement is characterized by its probability distribution and consequently by its statistical ensemble.

Thus, a statistical set (ensemble) in quantum theory is a set of identical measurements (experiments) made on an object in a given quantum state. To one and the same quantum state correspond many different ensembles, corresponding to various types of measurements. To specify statistical ensemble in quantum theory we must first determine the type of the measurement to be made on the object, and secondly give the state of the object. Thus, ρ_{nn} is the probability distribution of measurements of quantities A_n made in the state ρ. The probability distribution of measurement of another quantity B is characterized by the diagonal elements of ρ in B-representation

$$B\Psi'_n = b_n \Psi'_n \,, \quad \Psi'_n = U\Psi_n$$

$$\rho'_{nn} = \sum_{km} (U^{-1})_{nk}\, \rho_{km}\, U_{mn}$$

In the same way, we can find out the probability distribution in any other ensemble of measurements corresponding to the state ρ.

Up till now we did not use the notion of the wavefunction. As is known, originally quantum mechanics was based on this notion and the density matrix was introduced later. Now we will show that wave functions describe a very specific kind of quantum states, namely, the most ordered (non-chaotic) states - so called pure states.

To talk about statistical properties of the state, we should have some measure of these properties. Such a measure is provided to us by the information theory and is called the entropy of an ensemble [1]. (Of course, this entropy is a generalization of the entropy used in statistical mechanics).

The entropy of the ensemble is a measure of the statistical scatter or chaotic nature of the probability distribution in the ensemble. By definition, the entropy satisfies the following conditions. It is a functional of the probability distribution which has its maximum values in the most chaotic ensembles in which all members of the ensemble are found with equal probability. The entropy has its minimum value (zero) when the measured quantity has a definite value, i.e. all members of the ensemble have this definite value with probability equal to unity. And, lastly, the entropy must be additive: the entropy of a system consisting of two statistically independent subsystems is equal to the sum of the entropies of each subsystem. All these conditions (except for the inclusion of a constant factor) are satisfied by the quantity

$$\mathcal{E} = -\sum_i{}' p_i \ln p_i \tag{1.14}$$

where p_i is the probability with which i-th term of the statistical ensemble appears ($\sum_i p_i = 1$).

Now let us return to the statistical ensembles corresponding to some quantum state with the density matrix ρ. There are many ensembles corresponding to this state but we always can find out the less chaotic ensemble - i.e. with minimal entropy.

This may be done with the aid of the mathematical theorem that states: in the representation in which ρ is diagonal, the sum

$$\sum_n{}' \rho_{nn} \ln \rho_{nn} \qquad (\rho_{nm} = 0\,, \; n \neq m)$$

is larger than in any other representation with $\rho'_{nm} \neq 0$ $(n \neq m)$

$$\sum_n{}' \rho_{nn} \ln \rho_{nn} > \sum_n{}' \rho'_{nn} \ln \rho'_{nn}$$

Thus the quantity

$$\mathcal{E} = -\sum_n{}' \rho_{nn} \ln \rho_{nn} \equiv -\mathrm{Tr}(\rho \ln \rho) \tag{1.15}$$

may serve as a measure of the statistical properties of the state ρ: it is the entropy of the most ordered ensemble (with minimal entropy) of the state ρ.

Now, it is natural to distinguish the special class of the states with the entropy equal zero. These are the most ordered states. It is easy to see that for these states, ρ_{nm}, in the representation in which it is diagonal, is equal

$$\rho_{nm} = \delta_{n_0 n}\, \delta_{n_0 m} \tag{1.16}$$

in the arbitrary representation

$$\rho^2 = \rho\,, \tag{1.17}$$

which together with the equation

$$\mathrm{Tr}(\rho \ln \rho) = 0$$

characterize the class of pure states. From (1.16) follows an important property of the density matrix of pure states. In the arbitrary representation, it may be factorized

$$\rho_{nm} = c'_n c'^{*}_m \tag{1.18}$$

To prove this property it is enough to use (1.11) and transform (1.16) to the arbitrary representation

$$\rho'_{nm} = (U^{-1})_{nn_0} U_{n_0 m}$$

and to designate

$$(U^{-1})_{nn_0} = c'_n$$

Now the pure state may be characterized by the set of numbers C'_n . This set of numbers is usually called the wavevector. The mean value of the arbitrary operator A may be determined by the wavevector according to the formulae (1.3) and (1.18)

$$\langle A \rangle = \sum_{mn} \rho_{nm} A_{mn} = \sum_{mn} c^{*}_m A_{mn} c_n = \langle c | A | c \rangle \tag{1.19}$$

In particular when indices n take on a continuous set of meanings q (coordinates of particles) we call C_m a wavefunction, designate it as $C_n \to \Psi(q)$ and instead of summation in the formula (1.19) we use the integration

$$\langle A \rangle = \int \Psi^{*}(q) A \Psi(q) dq \ , \quad \int \Psi^{*}(q) \Psi(q) dq = 1 \tag{1.20}$$

The density matrix in this representation is equal

$$\rho(q,q') = \Psi(q) \Psi^{*}(q')$$

and according to the general rule the probability distribution takes the form

$$P(q) = \rho(q,q) = |\Psi(q)|^2$$

The usual introduction of the density matrix is based on this formula and on its expansion in some full set of the eigenfunctions

$$\Psi(q) = \sum_m a_m \Psi_m \tag{1.21}$$

$$P(q) = \sum a_n a_m^* \Psi_m^* \Psi_n \tag{1.22}$$

Then this description is generalized to the case when it is impossible to factorize the coefficients of the expansion (1.22)

$$P(q) = \sum \rho_{nm} \Psi_m^*(q) \Psi_n(q) =$$

$$= \sum \rho_{nn} |\Psi_n(q)|^2 + \sum_{n \neq m} \rho_{nm} \Psi_n(q) \Psi_m^*(q)$$

In the particular case when $\Psi_n(q)$ are the eigenfunctions of the momentum operator, ρ_{nm} characterize the coherent properties of the diffraction picture. The maximum coherence is achieved when $\rho_{nm} = a_n^* a_m$ i.e. in the pure state. Thus, in a sense, the pure state is the most coherent state.

4. The basic equation describing time evolution of quantum systems is the von Neumann equation

$$i\hbar \frac{\partial \rho}{\partial t} = [H, \rho] \tag{1.23}$$

where H is the Hamiltonian of the system. For pure states (1.18), the Schrödinger equation follows from the equation (1.23)

$$(i\hbar \dot{c}_n - \sum_k H_{nk}) c_m^* + (i\hbar \dot{c}_m^* + \sum_k H_{mk}^* c_k^*) c_n = 0$$

$$i\hbar \dot{c}_n = \sum_k H_{nk} c_k \quad \text{or} \quad i\hbar \dot{c} = Hc$$

and in q-representation (1.21)

$$i\hbar \frac{\partial \Psi}{\partial t} = H\Psi \tag{1.24}$$

When equations (1.23) or (1.24) are used to describe a change of quantum state, it is assumed that the operators of the physical quantities are not time-dependent, and the whole time-dependence is contained in the density matrix. This description or representation is named after Schroedinger. The time-dependence of the density matrix in this representation can be written in the form

$$\rho(t) = \exp(-iHt/\hbar)\,\rho(0)\,\exp(iHt/\hbar) \tag{1.25}$$

This equation is easily verified to satisfy (1.23). Another way of describing the change in a quantum state is the Heisenberg representation. In this representation all the time-dependence is transferred to the operators and the density matrix is time-independent. Both these representations are, of course, physically equivalent. For example, when taking mean

$$\langle A(t)\rangle = \mathrm{Tr}\,(\rho(t)A) = \mathrm{Tr}\,(\rho A(t)) \tag{1.26}$$

the same time function is obtained no matter which is time-dependent - the density matrix or the operator A. From (1.25) and (1.26) it follows

$$A(t) = \exp(iHt/\hbar)\,A\exp(-iHt/\hbar) \equiv U^{-1}(t)\,A\,U(t) \tag{1.27}$$

where the unitary transformation is defined by the operator

$$U(t) = \exp(-iHt/\hbar) \tag{1.28}$$

From (1.28) it follows that in the Heisenberg representation the operators satisfy the equations of motion

$$\frac{dA}{dt} = \frac{i}{\hbar}[H,A]\,, \qquad \frac{dH}{dt} = 0 \tag{1.29}$$

1.2 Two simple quantum-mechanical systems

This part is devoted to two simple quantum-mechanical systems, which may be described quite comprehensively and without approximations, i.e., exactly.

One of these systems is the two-state system and the other is a harmonic oscillator. Both of these systems have widespread physical applications and they constitute the essential part of the quantum mechanical equipment of our course.

(1) Two-state systems

Among the simplest quantum mechanical systems are those having only two possible states. There are many examples of physical systems with two states (or situations where only two states of the system are important).

A particle with spin $\frac{1}{2}$ provides the most known example of such a situation. In this case, the projection of the spin onto z-axis may take up two values: $\frac{1}{2}$ and $-\frac{1}{2}$; the energy of particles of this kind in a magnetic field also takes up two values. There are, accordingly, two eigenfunctions which describe the states with the projection of the spin: $\frac{1}{2}$ and $-\frac{1}{2}$.

There are many situations where only two levels of the molecules are important for the problem.

The nucleon may be in two states, i.e., proton and neutron.

The ammonia molecule has two states with equal energy and different configurations, which cannot be transformed one into the other by means of translations and rotations only. One is the "mirror" reflection of another. Chemical physics provides us with many other examples of this kind.

How to describe such two-state systems? To answer this question, it is worthwhile to recall the description of the particle with spin $\frac{1}{2}$ in the magnetic field $\vec{H}$. Spin operator $\vec{I}$ describes the angular momentum of the elementary particle (electron, neutron). In units of $\hbar$ it is expressed through the Pauli operators σ_x, σ_y, σ_z

$$\vec{I} = \frac{1}{2}\vec{\sigma}$$

The Pauli operators $\vec{\sigma}$ in the representation where σ_z is diagonal are equal

$$\sigma_x = \begin{pmatrix} 0 & 1 \\ 1 & 0 \end{pmatrix}, \quad \sigma_y = \begin{pmatrix} 0 & -i \\ i & 0 \end{pmatrix}, \quad \sigma_z = \begin{pmatrix} 1 & 0 \\ 0 & -1 \end{pmatrix} .$$

The magnetic moment of the particle is proportional to its spin

$$\vec{M} = \gamma \vec{I}$$

where γ is the so-called gyromagnetic ratio. In the external magnetic field $\vec{H}$ the energy of the spin, i.e., its Hamiltonian has the form

$$\mathcal{H} = -\vec{M}\vec{H} = -\gamma\vec{I}\vec{H} \tag{1.30}$$

If we choose the coordinate axis z in the direction of the magnetic field, we will find for the energy eigenvalues the following expressions

$$E_{-1/2} = -\frac{1}{2}\gamma H_z , \quad E_{1/2} = \frac{1}{2}\gamma H_z$$

Now, using general quantum mechanical rule (1.29)

$$\frac{dA}{dt} = \frac{i}{\hbar}[\mathcal{H}, A]$$

and the commutation relations for the spin operators

$$[I_x, I_y] = iI_z, \quad [I_y, I_z] = iI_x, \quad [I_z, I_x] = iI_y$$

we find from (1.30) that the equations of motion for spin operators and for the magnetic moment have the form

$$\dot{\vec{I}} = \gamma \vec{I} \times \vec{H}, \quad \dot{\vec{M}} = \gamma \vec{M} \times \vec{H} \tag{1.31}$$

These are the equations describing the time behavior of the operators $\vec{I}$ and $\vec{M}$. From equation (1.31), the equations for mean values of $\vec{I}$ and $\vec{M}$ may be easily derived. For this purpose, the following property of the operator of the derivative operator $\dot{A}$ should be utilized

$$\frac{d\langle A\rangle}{dt} = \langle \dot{A}\rangle \tag{1.32}$$

The proof of this property is very simple:

$$\frac{d}{dt}\langle A\rangle = \frac{d}{dt}\mathrm{Tr}\left(\rho\, e^{iHt/\hbar} A\, e^{-iHt/\hbar}\right) = \mathrm{Tr}\left(\rho \frac{i}{\hbar}[H, A]\right) = \langle \dot{A}\rangle$$

Thus, from Eqs. (1.31) and (1.32) it follows that the equations for the mean values of $\vec{I}$ and $\vec{M}$ coincide with the equations (1.31) for the operators. These equations describe the precession of the magnetic moment $\vec{M}$ in the magnetic field $\vec{H}$. If, for example, the constant magnetic field $\vec{H}$ is directed along the z-axis, then the equations of motion take on a very simple form

$$\dot{M}_x = -\omega_0 M_y, \quad \dot{M}_y = \omega_0 M_x, \quad \dot{M}_z = 0$$

$$\ddot{M}_x + \omega_0^2 M_x = \ddot{M}_y + \omega_0^2 M_y = 0, \quad \omega_0 = \gamma H_z/\hbar$$

Now, let us perform the generalization to the arbitrary two-state systems. We introduce the effective spin operator $\vec{r}$ defined by its three components, r_1, r_2, r_3. The operator $\vec{r}$ is a vector not in an ordinary geometrical space but in some abstract space. The components of the vector $\vec{r}$ are defined by the commutation relations

$$\vec{r} \times \vec{r} = i\vec{r}\,, \quad [r_1, r_2] = i r_3\,, \quad [r_2, r_3] = i r_1\,, \quad [r_3, r_1] = i r_2 \tag{1.33}$$

and the equations

$$r_1^2 = r_2^2 = r_3^2 = \frac{1}{4} \tag{1.34}$$

In the representation in which r_3 is diagonal, the components r_i take the form

$$r_1 = \frac{1}{2}\begin{pmatrix} 0 & 1 \\ 1 & 0 \end{pmatrix}, \quad r_2 = \frac{1}{2}\begin{pmatrix} 0 & -i \\ i & 0 \end{pmatrix}, \quad r_3 = \frac{1}{2}\begin{pmatrix} 1 & 0 \\ 0 & -1 \end{pmatrix} \tag{1.35}$$

It is easy to verify that these operators (1.35) satisfy the relations (1.33,34). We will show that these effective spin operators can describe any two-state system regardless of its nature. For this purpose it is enough to realize that the group of the linear Hermitian operators r_1, r_2, r_3 and the unit operator

$$I = \begin{pmatrix} 1 & 0 \\ 0 & 1 \end{pmatrix}$$

have the following property:
Every linear Hermitian operator A relating to a two-state system may be expressed in terms of these operators r_1, r_2, r_3, I

$$\begin{aligned} A &= aI + br_1 + cr_2 + dr_3 = \\ &= \begin{pmatrix} a + \frac{1}{2}d & \frac{1}{2}(b - ic) \\ \frac{1}{2}(b + ic) & a - \frac{1}{2}d \end{pmatrix} \end{aligned} \tag{1.36}$$

The r.h.s. of (1.36) is just the representation of an arbitrary linear Hermitian operator of a quantity characterized by two states.

In particular, the Hamiltonian of the two-state system may be expressed through the operators of the effective spin

$$\mathcal{H} = E_0 I + K_1 r_1 + K_2 r_2 + K_3 r_3$$

Similarly, as for Eq. (1.31), we get from Eq. (1.37) the equations of motion for operators and their mean values

$$\vec{r} = -\frac{1}{\hbar}\left[\vec{r} \times \vec{K}\right] \qquad (1.37)$$

Now, let us turn to the eigenfunction problem and the transformation properties. Let Ψ_1 and Ψ_2 be two eigenfunctions of some two-state operator (e.g., of the Hamiltonian). In another representation

$$\psi_1' = a\psi_1 + b\psi_2 = U\psi_1\,, \quad \psi_2' = c\psi_1 + d\psi_2 = U\psi_2$$

The inverse transformation gives

$$\psi_1 = \frac{d\psi_1' - b\psi_2'}{ad - bc} = U^{-1}\psi_1'\,, \quad \psi_2 = \frac{-c\psi_1' + a\psi_2'}{ad - bc} = U^{-1}\psi_2'$$

Assuming normalization of all these functions to unity, we get

$$|a|^2 + |b|^2 = 1\,, \quad |c|^2 + |d|^2 = 1$$

$$|d|^2 + |b|^2 = \Delta^2\,, \quad |a|^2 + |c|^2 = \Delta^2\,, \quad \Delta = ad - bc = 1$$

Thus,

$$\psi_1' = U_{11}\psi_1 + U_{21}\psi_2\,, \quad \psi_2' = U_{12}\psi_1 + U_{22}\psi_2$$

$$\psi_1 = (U^{-1})_{11}\psi_1' + (U^{-1})_{21}\psi_2'\,, \quad \psi_2' = (U^{-1})_{12}\psi_1' + (U^{-1})_{22}\psi_2'$$

Having in mind the unitary property of the operator U

$$U^{+} = U^{-1} \quad \text{or} \quad U_{ik}^{*} = (U^{-1})_{ki}$$

we get

$$a = d^{*} \quad , \quad b = -c^{*}$$

Thus, in general, the unitary transformation for a two-state has the form

$$\psi_1' = a\psi_1 + b\psi_2 \quad , \quad \psi_2' = -b^{*}\psi_1 + a^{*}\psi_2 \tag{1.39}$$

The eigenfunctions of the effective spin may be cast in a very simple form. Let Ψ_1 and Ψ_2 be the eigenfunctions of spin operator r_3

$$r_3 \psi_1 = \frac{1}{2}\psi_1 \quad , \quad r_3 \psi_2 = -\frac{1}{2}\psi_2$$

For the two-state system, the arbitrary wave vector, i.e., the representation of the wave functions by the expansion coefficients C_n

$$\psi = \sum_n c_n \varphi_n$$

has only two components and can be represented as

$$c = \begin{pmatrix} c_1 \\ c_2 \end{pmatrix}$$

Then, it is easy to verify that $\Psi_1 = \begin{pmatrix} 1 \\ 0 \end{pmatrix}$; $\Psi_2 = \begin{pmatrix} 0 \\ 1 \end{pmatrix}$ are the eigenfunctions of the operator r_3 normalized to unity.

The problem of the time evolution of the density matrix of the two-state system may be solved exactly as well.

Let us consider the system characterized by two energy levels E_1 and E_2 and the perturbation energy V giving rise to transitions between these levels. Then, the Hamiltonian of such a two-level system may be written in the form

$$\mathcal{H} = n_1 E_1 + n_2 E_2 + r_+ V_{12} + r_- V_{21} \tag{1.40}$$

where the operators n_1, n_2, $r_\pm$ are expressed through the operators of the effective spin:

$$n_1 = \frac{1}{2} - r_3 \; , \quad n_2 = \frac{1}{2} + r_3 \; , \quad r_\pm = r_1 \pm i r_2 \tag{1.41}$$

It is clear that n_1, n_2 have the meaning of the occupation number operators of the states 1 and 2:

$$\begin{aligned} n_1 = 1,\ n_2 = 0 \quad &\text{if} \quad r_3 = -\tfrac{1}{2} \\ n_1 = 0,\ n_2 = 1 \quad &\text{if} \quad r_3 = \tfrac{1}{2} \end{aligned} \tag{1.42}$$

For simplicity we would consider the case when

$$V_{12} = V_{21} = V \tag{1.43}$$

In this case the Hamiltonian (1.40) may be written in the form

$$\mathcal{H} = (E_1 + E_2)/2 + r_3 (E_2 - E_1) + 2 r_1 V \tag{1.44}$$

Then, time evolution of the density matrix is determined by the expression (see (1.25))

$$\begin{aligned} \rho(t+\tau) &= e^{-i\tau\mathcal{H}/\hbar} \rho(t) e^{i\tau\mathcal{H}/\hbar} = R(\tau)\rho(t) = \\ &= \sum_{m'n'} R(\tau)_{mn}^{m'n'} \rho(t)_{m'n'} \end{aligned} \tag{1.45}$$

As has been mentioned above, an arbitrary function relating to the two-state system may be represented as a linear combination of the effective spin operators. The corresponding formula has the form [2]

$$f(a + \vec{b}\vec{r}) = A + \vec{B}\vec{r} \tag{1.46}$$

where

$$\begin{aligned} A &= \frac{1}{2}\left[f(a + \tfrac{1}{2}b) + f(a - \tfrac{1}{2}b)\right] \\ \vec{B} &= \frac{\vec{b}}{2b}\left[f(a + \tfrac{1}{2}b) - f(a - \tfrac{1}{2}b)\right] \end{aligned} \tag{1.47}$$

Thus

$$\begin{aligned} \exp(-i\tau\mathcal{H}/\hbar) &= \exp(-i\tau(E_1 + E_2)/2) \times \\ &\times \left[\cos(\Omega\tau/2) - 2i\sin(\Omega\tau/2)(b_1 r_1 + b_3 r_3)\right] \end{aligned} \tag{1.48}$$

where

$$\Omega = \frac{1}{\hbar}\left[(E_2 - E_1)^2 + 4V^2\right]^{1/2} \tag{1.49}$$

$$b_3 = (E_2 - E_1)/\hbar\Omega \;,\quad b_1 = 2V/\hbar\Omega \;,\quad b_1^2 + b_3^2 = 1$$

The density matrix of the two-state system may be presented in the form

$$\rho = n_1\rho_{11} + n_2\rho_{22} + r_+\rho_{12} + r_-\rho_{21} \tag{1.50}$$

This representation corresponds to the eigenfunctions of the operators $n_1 n_2$ (or the operator r_3).

Substituting the expressions (1.50) and (1.48) into (1.45) we get for the matrix elements $R_{mn}^{m'n'}$ of the supermatrix R (i.e., the matrix transforming matrix $\rho(t)$ into another matrix $\rho(t+\tau)$) the following expressions

$$
\begin{aligned}
R_{11}^{11} &= \tfrac{1}{2} + \tfrac{1}{2}\left[\cos^2(\tfrac{1}{2}\Omega\tau) + \sin^2(\tfrac{1}{2}\Omega\tau)(b_3^2 - b_1^2)\right] = R_{22}^{22} \\
R_{22}^{11} &= \tfrac{1}{2} - \tfrac{1}{2}\left[\cos^2(\tfrac{1}{2}\Omega\tau) + \sin^2(\tfrac{1}{2}\Omega\tau)(b_3^2 - b_1^2)\right] = R_{11}^{22} \\
R_{12}^{11} &= -\left[i\cos(\tfrac{1}{2}\Omega\tau)\sin(\tfrac{1}{2}\Omega\tau b_1) - \sin^2(\tfrac{1}{2}\Omega\tau) b_3 b_1\right] = R_{11}^{12} \\
R_{21}^{11} &= \left[i\cos(\tfrac{1}{2}\Omega\tau)\sin(\tfrac{1}{2}\Omega\tau b_1) + \sin^2(\tfrac{1}{2}\Omega\tau) b_3 b_1\right] = R_{11}^{21} \\
R_{12}^{22} &= \left[i\cos(\tfrac{1}{2}\Omega\tau)\sin(\tfrac{1}{2}\Omega\tau b_1) - \sin^2(\tfrac{1}{2}\Omega\tau) b_3 b_1\right] = R_{22}^{12} \\
R_{21}^{22} &= -\left[i\cos(\tfrac{1}{2}\Omega\tau)\sin(\tfrac{1}{2}\Omega\tau b_1) + \sin^2(\tfrac{1}{2}\Omega\tau) b_3 b_1\right] \\
R_{12}^{12} &= \left[\cos^2(\tfrac{1}{2}\Omega\tau) + 2i\cos(\tfrac{1}{2}\Omega\tau)\sin(\tfrac{1}{2}\Omega\tau) b_3 - \sin^2(\tfrac{1}{2}\Omega\tau) b_3\right] = R_{21}^{21\,*} \\
R_{21}^{12} &= \sin^2(\tfrac{1}{2}\Omega\tau b_1) = R_{12}^{21}. \\
R_{22}^{21} &= -R_{11}^{21}
\end{aligned}
\tag{1.51}
$$

These formulas and the expression (1.45) solve the problem of time evolution of the density matrix of a two-state system in the representation in which n_1, n_2 or r_3 are diagonal.

(2) Harmonic oscillator

Another simple quantum-mechanical system widely used in various areas of physics and chemistry is the harmonic oscillator. The harmonic oscillator in classical mechanics is characterized by its Hamiltonian function (i.e., the energy expressed through canonical coordinate q and momentum p)

$$\mathcal{H} = \frac{p^2}{2m} + \frac{m\omega^2 q^2}{2}$$

Later on, we will choose the units with m=1. In quantum mechanics the harmonical oscillator is characterized by its Hamiltonian

$$\mathcal{H} = \frac{1}{2} p^2 + \frac{1}{2} \omega^2 q^2 \tag{1.52}$$

where the operators of the coordinate q and the momentum p in q-representation take the form: q is the operator of multiplication and

$$p = -i\hbar \frac{\partial}{\partial q}$$

These operators satisfy the commutation relation

$$[q, p] = i\hbar \tag{1.53}$$

From Eqs. (1.52) and (1.53) and the general quantum mechanics formula (1.29)

$$\dot{A} = \frac{i}{\hbar} [\mathcal{H}, A]$$

it follows that the equations of motion have the form

$$\dot{p} = -\omega^2 q^2 , \quad \dot{q} = p$$

$$\ddot{q} + \omega^2 q = 0$$

From these equations and the relation (1.32), we find that the equations for the mean values of q and p coincide with those for operators and also with equations for the classical coordinate and momentum.

Then in every course of quantum mechanics it is shown that the eigenvalues of the energy are equal

$$E_n = \left(n+\frac{1}{2}\right)\hbar\omega \qquad n = 0,1,2,\ldots \tag{1.54}$$

that in the energy representation matrix elements q and p are

$$\begin{aligned} q_{n,n+1} &= q^{*}_{n+1,n} = \left[\hbar(n+1)/2\omega\right]^{1/2} \\ p_{n,n+1} &= p^{*}_{n+1,n} = -i\left[\hbar\omega(n+1)/2\right]^{1/2} \end{aligned} \tag{1.55}$$

and the eigenfunctions are equal

$$\Psi_n(q) = \left(\frac{\omega}{\pi\hbar}\right)^{1/4}\left(2^n n!\right)^{-1/2}\exp\left(-\omega^2 q^2/2\hbar\right) H_n\left(q(\omega/\hbar)^{1/2}\right)$$

where $H_n(x)$ is the n-th Hermitian polynomial. Sometimes it is convenient to introduce the operators a and a^+, which are called the "annihilation" and "creation" operators, respectively

$$q = (\hbar/2\omega)^{1/2}(a+a^+), \qquad p = i(\hbar\omega/2)^{1/2}(a^+-a) \tag{1.56}$$

By using the commutation relation (1.53), it is easy to find that the operators a and a^+ are subject to the following commutation relation

$$[a, a^+] = 1 \tag{1.57}$$

We notice that the operators a and a^+ are non-Hermitian unlike those of q and p. The operators a and a^+ can be used in order to write the energy of the oscillator in the form

$$\mathcal{H} = \frac{1}{2}\hbar\omega\left(a^+a + aa^+\right) = \hbar\omega\left(n+\frac{1}{2}\right) \tag{1.58}$$

where the symbol n denotes the particle-number operator

$$n = a^{+}a \tag{1.59}$$

which takes non-negative integer eigenvalues

$$n = 0, 1, 2, \ldots$$

For the quantities a and a^{+} (both for the operators and for mean values) we get extremely simple equations

$$\dot{a} = -i\omega a \;, \qquad \dot{a}^{+} = i\omega a^{+} \tag{1.60}$$

with solutions

$$a(t) = a(0)e^{-i\omega t} \qquad\qquad a^{+}(t) = a^{+}(0)\,e^{i\omega t}$$

The operators a and a^{+} have very simple matrix elements

$$a_{n,n+1} = a^{+}_{n+1,n} = (n+1)^{1/2} \;, \tag{1.61}$$

and their action on the eigenfunctions is determined by the expressions

$$a\,\Psi_{n+1} = (n+1)^{1/2}\,\Psi_{n} \;, \qquad a^{+}\Psi_{n} = (n+1)^{1/2}\,\Psi_{n+1}$$

The meaning of the names "annihilation" and "creation" operators may be understood from these relations.

1.3 Transition probability per unit time. Master equations. General properties of irreversible motion.

Here we want to analyze a notion which is very important in the theory of rate processes, the transition probability per unit time.

Let us consider the quantum-mechanical system described by the Hamiltonian

$$\mathcal{H} = \mathcal{H}_0 + V \tag{1.62}$$

where V is part of the whole Hamiltonian $\mathcal{H}$ which may be considered as a small perturbation. The meaning of the word "small" would be clear from further derivation. We would consider the transition between eigenstates of the unperturbed Hamiltonian which are caused by the perturbation energy V.

According to general principles of quantum mechanics, time evolution of the quantum system is described by the von Neumann equation (1.23)

$$i\hbar \frac{\partial \rho}{\partial t} = [\mathcal{H}_0 + V, \rho] \tag{1.63}$$

When V may be considered as a small perturbation, it is convenient to use so-called interaction representation. This representation is, in a sense, intermediate between the Heisenberg and Schroedinger representations. In the interaction representation the time dependence of the operators is determined by the unperturbed Hamiltonian

$$A_{int} = U_0^{-1} A U_0 \ , \quad U_0 = \exp\left(-\frac{i}{\hbar}\mathcal{H}_0 t\right)$$
$$V_{int} = U_0^{-1} V U_0 \tag{1.64}$$

The density matrix in this representation

$$\rho_{int} = U_0^{-1} \rho U_0 \tag{1.65}$$

satisfies, as it is easy to show from (1.63)-(1.65), the equation

$$i\hbar \frac{\partial \rho}{\partial t} = [V, \rho] \tag{1.66}$$

Here and further on we will omit indice "int".

We start from the equation (1.66) and consider V to be a small quantity. We expand the density matrix ρ into a series

$$\rho = \sum_k \rho^{(k)} \tag{1.67}$$

where $\rho^{(k)}$ is a quantity of the k-th order of smallness with respect to V. Substituting (1.67) for (1.66) we obtain the system of the recurrence relations

$$i\hbar \frac{\partial \rho^{(k)}}{\partial t} = [V, \rho^{(k-1)}] \tag{1.68}$$

As the zero approximation, we shall take the value of the density matrix at $t = 0$. We write the density matrix with an accuracy up to terms of the second order of smallness

$$\rho(t) = \rho^{(0)} + \rho^{(1)} + \rho^{(2)} + \ldots = \rho(0) - \frac{i}{\hbar}\int_0^t dt_1 \left[V(t_1), \rho(0)\right]$$

$$- \frac{1}{\hbar^2}\int_0^t dt_1 \int_0^{t_1} dt_2 \left[V(t_1), \left[V(t_2), \rho(0)\right]\right] + \ldots \tag{1.69}$$

Now, let us examine the special case when initially the system was known to be in the stationary state n_o (with definite energy E_{n_o}) of the unperturbated Hamiltonian

$$\rho_{mn}(0) = \delta_{mn_0}\,\delta_{nn_0} \tag{1.70}$$

Substituting this expression into the r.h.s. of (1.69), we find

$$\rho_{nn}(t) = \frac{2\left|V_{nn_0}\right|^2}{\left(E_{n_0} - E_n\right)^2}\left[1 - \cos\left(\left(E_{n_0} - E_n\right)t/\hbar\right)\right], \quad n \neq n_0 \tag{1.71}$$

Here E_n and E_{no} are the energy levels of the unperturbed system. Thus, $\rho_{nn}(t)$ has the meaning of the probability of transition from the state n_o, where the system was at the moment of time $t=0$, to the state n.

Now we shall derive the expression for transition probability per unit time. This expression plays a central role in the theory of rate processes. That is why we would try to explain in more detail the meaning and limitations of this expression.

The energy levels of the initial and final states often form a continuous spectrum. Let us examine in greater detail the case (which is of physical interest for the theory of rate processes) when the energy of the final state is a part of the continuous spectrum.

We shall assume that the variables $\underset{\sim}{n}$ that describe the final state consist of E (the energy) and a certain set of variables (including the continuous set as well) which we shall denote by the suffix $\underset{\sim}{u}$. Since by assumption, E varies continuously, we can introduce a number of states with fixed values u in the energy range dE

$$dZ_u = \rho_u(E)\,dE \tag{1.72}$$

We are generally interested in the probability of a transition into a state with a fixed value u but with any value E (of course E will be probably near to energy E_o of the initial state). The required probability becomes

$$W_{n_o u}(t) = \sum_{E_n,\, n \neq n_o} \rho_{nn}(t) =$$

$$= \sum_{E_u,\, u \neq n_o} \frac{2|V_{n_o u}|^2}{(E_u - E_{n_o})^2} \left[1 - \cos\left((E_{n_o} - E_u)t/\hbar\right)\right] \tag{1.73}$$

$$= \int dE\, \rho_u(E) 2 |\langle E_o, n_o|V|E, u\rangle|^2 (E - E_o)^{-2} \left[1 - \cos\left((E_o - E)t/\hbar\right)\right]$$

Now, under certain limitations, this expression is proportional to time, and then we are able to speak about the transition probability per unit time. Again, because of the importance of this notion, "transition probability per unit time", and frequent misunderstanding of its limitations, we will try to analyze them in detail.

Let us introduce a new variable of integration

$$y = \frac{1}{2\hbar}(E - E_o)t, \quad E = E_o + 2\hbar y/t, \quad dy = \frac{t}{2\hbar}\, dE$$

then the expression (1.73) may be rewritten as

$$\bar{W}_{n_o u} = \frac{2\pi}{\hbar} t \int_{-(E_o - E_{min})t/2\hbar}^{(E_{max} - E_o)t/2\hbar} dy\, f(E_o + 2\hbar y/t)\, \frac{\sin^2 y}{\pi y^2} \tag{1.74}$$

where

$$f(E) = |\langle E_o, u_o|V|E, u\rangle|^2 \rho_u(E)$$

and E_{min}, E_{max} are minimal and maximal meanings of the energy spectrum of the system. Now let

$$\Delta E = \hbar\omega^*$$

denote the range of E near E_o in which the function f(E) changes only slightly when E changes by an amount which is small compared with $\Delta E = \hbar\omega^*$. In other words, $\hbar\omega^*$ is the characteristic scale of the variation of the quantity E near E_o. Then, let us suppose that time t is so large that the following inequalities are satisfied

$$t \gg (\omega^*)^{-1},\ \hbar(E_0 - E_{min})^{-1},\ \hbar(E_{max} - E_0)^{-1} \tag{1.75}$$

We notice further that the function $\sin^2 y/\pi y^2$ gives its main contribution to the integral in the region $-\pi < y < \pi$ (i.e., for values $y^2 \sim 1$). By using this fact, the definition of ω^* and the inequalities (1.75) we get

$$\bar{W}_{n_0 u} = \frac{2\pi}{\hbar} t f(E_0) \int_{-\infty}^{\infty} dy \frac{\sin^2 y}{\pi y^2} = \frac{2\pi}{\hbar} f(E_0) t \tag{1.76}$$

or

$$\bar{W}_{n_0 u} = \frac{2\pi}{\hbar} |\langle E_0 n_0 | V | E_0 u\rangle|^2 \varrho_u(E_0) t = w_{n_0 u} t \tag{1.77}$$

Thus, under conditions (1.75), i.e., for a large enough t, the transition probability is proportional to time t and it is possible to define the transition probability per unit time $w_{n_0 u}$ (which itself does not depend on time). Of course, the quantities determining the condition (1.75): ω^*, E_{min} and E_{max} are defined by the actual properties of the system under investigation. It should be stressed that the applicability of (1.77) depends not only on condition (1.75); we must also satisfy the condition concerning the smallness of t compared with the characteristic time required for significant change to occur in the state of the system. This is the condition of applicability of the perturbation theory. In particular

$$w_{n_0 u} t \ll 1 \tag{1.78}$$

should be satisfied. For both these conditions (1.75) and (1.78) to be satisfied, it is necessary

$$\omega^* \gg w_{n_0 u} \tag{1.79}$$

This condition determines the limitation of the perturbation theory ($w_{n_0 u}$ proportional to $|V_{nm}|^2$).

It should be mentioned also that for small values of time

$$t \ll (\omega^*)^{-1}$$

there is no time-independent transition probability per unit time. In this case, according to (1.73) the transition probability is proportional to t^2.

Very frequently the transition probability per unit time is expressed in the form

$$w_{nu_0} = \frac{2\pi}{\hbar} |V_{n_0 u}|^2 \delta(E - E_0) \tag{1.80}$$

It should, however, be remembered that this expression has meaning only after integration over the energy. Then it obtains the form (1.77)

$$w_{n_0 u} = \frac{2\pi}{\hbar} |\langle E_0 n_0 | V | E_0 u \rangle|^2 \rho_u (E_0) \tag{1.81}$$

Of course, in reality we deal with quasi-continuous spectra of energy. The question arises when, under what conditions, the quasi-continuous spectrum, characterized by some mean energy interval δE between energy levels, may be considered as a continuum. The inspection of transition from summation to integration in (1.73) shows that this transition is valid, provided

$$t \ll \hbar/\delta E$$

This means that δE should satisfy the condition

$$\omega^* \gg w_{n_0 u} \gg \delta E/\hbar \tag{1.82}$$

since $w^{-1}_{n_0 u}$ is the characteristic time scale of the problem. The formula (1.81) (or the equivalent formula (1.82) is sometimes called the Fermi's Golden Rule Formula. This formula was served by Pauli [3] as a basis of his, to a certain extent, intuitive derivation of master equations (or Pauli equation). These equations govern the time behavior of the diagonal elements of the density matrix

$$\dot{\rho}_{nn} = -\sum_k (w_{nk} \rho_{nn} - w_{kn} \rho_{kk}) \tag{1.83}$$

where w_{nk} are the transition probabilities between states n and k and $\rho_{kk}(t)$ is the probability of finding the system in the state k at the moment t. In the state of thermal equilibrium, the r.h.s. of (1.83) should vanish and ρ_{kk} should satisfy the Boltzmann relation

$$\rho_{nn}/\rho_{kk} = e^{-(E_n - E_k)/k_B T} \tag{1.84}$$

where E_k is the energy level of the state k. This implies the relation between transition probabilities

$$w_{nk} = w_{kn} \exp\left[-(E_n - E_k)/k_B T\right] \tag{1.85}$$

The equation (1.83) in contrast to the von Neumann equation (1.23) has irreversible character. Later on (see Chapter III) we will present rigorous derivation of the equations describing irreversible motion from the von Neumann equation.

Now we will prove that the irreversibility is connected with continuous spectrum of energies (or more exactly, with continuous spectrum of differences of energy eigenvalues). We will show that if $\langle A(t)\rangle$ is the mean value of a certain quantity at time t, then in a system whose differences of energy levels form a continuous specturm, $\langle A(t)\rangle$ approaches a definite limit when $t \to \infty$, provided general assumptions are made. The mean value of the quantity A at the moment t is

$$\langle A(t)\rangle = Tr(\rho(t) A)$$

and according to (1.25) it may be cast in the form

$$\langle A(t)\rangle = \sum_{n,m,u,u'} \rho_{nu,mu'}(0) A_{mu',nu} e^{-i\omega_{mn}t} = \int_{-\infty}^{\infty} d\omega\, g(\omega) e^{-i\omega t} \tag{1.86}$$

where n,m are suffixes denoting energy levels; u,u' are other quantum numbers; and $\omega_{nm} = (E_n - E_m)/\hbar$. The quantity $g(\omega)$, as can easily be checked, is defined by

$$g(\omega) = \sum_{u,u',n,m} \rho_{nu,mu'}(0) A_{mu',nu}\, \delta\left((E_n - E_m)/\hbar - \omega\right) \tag{1.87}$$

If the system has a discrete energy spectrum, $g(\omega)$ is a discontinuous function, A(t) is equal to the discrete sum of the harmonic functions and has no limit when $t \to \infty$. After a finite time the system will approach, as closely as one likes, to the original state.

For the system with a continuous spectrum, for $\langle A(t)\rangle$ to have a limit when $t \to \infty$ (which is an obvious condition of the irreversibility) it is sufficient for the function $g(\omega)$ to have the form

$$g(\omega) = G\delta(\omega) + h(\omega)$$

where $h(\omega)$ has no δ-type singularities and is absolutely integrable in the range $[-\infty, +\infty]$. (The appearance of the term $G\delta(\omega)$ is due to the diagonal terms $\rho_{nu;nu'}$). Then on the basis of the Lebesque-Riman theorem

$$\lim_{t\to\infty} \langle A(t)\rangle = G + \lim_{t\to\infty} \int_{-\infty}^{\infty} d\omega\, h(\omega) e^{-i\omega t} = G$$

From the above it is clear that the asymptotic value of $\langle A(\infty)\rangle$ is of the form

$$\langle A(\infty)\rangle = G = \sum_{nuu'} \rho_{nunu'}(0) A_{nu'nu}$$

It must be stressed that the proof given here is based essentially upon the assumption that $g(\omega)$ contains no δ-functions when $\omega \neq 0$. As can be seen from (1.86) this assumption relates both to the density matrix and to the operator which is of interest to us. In principle we can imagine an idealized situation when $g(\omega)$ contains δ-functions with $\omega \neq 0$. For example, if the matrix elements of A are non-zero only when $(E_n-E_m)/h = \pm\omega_o$ (as for the harmonic oscillator), then, as can be seen easily from (1.86)

$$g(\omega) = h_1 \delta(\omega - \omega_0) + h_2 \delta(\omega + \omega_0)$$

$$\langle A\rangle = h_1 \exp(i\omega_0 t) + h_2 \exp(-i\omega_0 t)$$

and $\langle A(t)\rangle$ has no limit when $t \to \infty$.

It is worthwhile to mention here that just in this case we cannot introduce the notion of the probability per unit time, which is also based on the assumption about continuous spectrum of energies.

1.4 The Born-Oppenheimer approximation. Phonons. Phonon-phonon interaction and relaxation

Chemical compounds, molecules - small and large (like proteins) as well as solids - all of them may be represented as a set of interacting electrons and nuclei. In the non-relativistic approximation the Hamiltonian of the system of electrons and nuclei may be written as

$$\mathcal{H} = T + H(q,\xi) \tag{1.88}$$

where T is the kinetic energy of the nuclei

$$T = \sum_k \frac{p_k^2}{2M_k} = -\frac{\hbar^2}{2}\sum_k \frac{1}{M_k}\frac{\partial^2}{\partial q_k^2}, \quad p_k \text{ and } q_k \quad \text{are}$$

the momenta and coordinates of the nuclei, M_k are their masses, and

$$H(q,\xi) = \sum_j \frac{\pi_j^2}{2m} + U(\xi,q) \tag{1.89}$$

is the electronic Hamiltonian which depends on nuclear coordinates q_k; π_j is the momentum of j-th electron, m is mass of the electron, ξ - coordinates of the electron and the term $U(\xi,q)$ represents the Coulomb interaction between electrons and nuclei.

Due to the fact that masses of electrons are much smaller than masses of nuclei, it is possible to utilize the so-called adiabatic approximation, or the Born-Oppenheimer approximation [6]. In this method we may approximately calculate the electronic eigen-functions, not taking into account the kinetic energy of nuclei, the latter being considered as a small perturbation. Describing the electronic motion, we assume that it is possible to neglect the motion of nuclei and that the electronic wavefunctions depend on nuclear coordinates as on parameters. Thus in zero approximation

$$H_0(q,\xi)\chi_\alpha(q,\xi) = E_\alpha(q)\chi_\alpha(q,\xi) \tag{1.90}$$

(The indice "0" shows that H_0 does not necessarily coincide with H. However, we usually assume that the difference $H-H_0$ is small). In this approximation we assume also that the nuclear wavefunctions satisfy the equation

$$\left(H_{\alpha\alpha}(q) + T_{\alpha\alpha}(q) + T\right)\Phi_{\alpha m}(q) = E_{\alpha m}\Phi_{\alpha m}(q) \tag{1.91}$$

where $H_{\alpha\alpha} = H_{\alpha\alpha}(q) + T_{\alpha\alpha}(q)$ is the Hamiltonian (1.88) of the system averaged over the electronic motions:

$$H_{\alpha\alpha}(q) = \int d\xi\, \chi_\alpha^* H(\xi, q)\, \chi_\alpha$$

$$T_{\alpha\alpha} = -\frac{\hbar^2}{2}\sum_k \frac{1}{M_k}\int d\xi\, \chi_\alpha^* \frac{\partial^2 \chi_\alpha}{\partial q_k^2} \qquad (1.92)$$

Thus, in this approximation the nuclear motion is described by the Hamiltonian depending only on nuclear coordinates and the role of the potential energy is played by the total Hamiltonian of the system, averaged over the electronic motion. The effective potential energy of nuclei is

$$U_\alpha(q) = H_{\alpha\alpha}(q) + T_{\alpha\alpha}(q) \qquad (1.93)$$

and it depends on the electronic state. In various electronic states there are various potential energies.

Now we should justify the assumptions (1.90) and (1.91), i.e., to show that the wavefunction of the total system, nuclei + electrons may be represented as

$$\Psi = \sum_{\alpha m} C_{\alpha m}\, \Phi_{\alpha m}(q)\, \chi_\alpha(q, \xi)\, e^{-iE_{\alpha m}t/\hbar} \qquad (1.94)$$

with coefficients $C_{\alpha n}$ which very slowly depend on time. To show the slowness of this dependence, we substitute (1.94) for the Schroedinger equation of the system

$$i\hbar \frac{\partial \Psi}{\partial t} = \mathcal{H}\Psi$$

$$i\hbar \sum_{\alpha m}\left[\dot{C}_{\alpha m}\Phi_{\alpha m}(q)\chi_\alpha(q,\xi) - \frac{iE_{\alpha m}}{\hbar} C_{\alpha m}\Phi_{\alpha m}(q)\chi_\alpha(q,\xi)\right] e^{-iE_{\alpha m}t/\hbar}$$

$$= \sum_{\alpha m} C_{\alpha m}\, \mathcal{H}\, \Phi_{\alpha m}(q)\, \chi_\alpha(q,\xi)\, e^{-iE_{\alpha m}t/\hbar}$$

We will multiply both parts of this equation by $\chi_\beta^*(q,\xi)\Phi_{\beta n}(q)$ and integrate over the variables ξ and q.

Using the orthonormality properties of these functions and (1.91) we get

$$i\hbar \dot{C}_{\alpha n} = \sum_{\beta} \langle n | H_{\beta\alpha}(1-\delta_{\alpha\beta}) + T_{\beta\alpha}(1-\delta_{\alpha\beta}) -$$
$$- \hbar^2 \sum_{j} \frac{1}{M_j} \left(\frac{\partial}{\partial q_j}\right)_{\beta\alpha} \frac{\partial}{\partial q_j} | m \rangle C_{\beta m} \exp\left(i(E_{\alpha m} - E_{\beta n})t/\hbar\right) \tag{1.95}$$

It is easy to see that the variation of $C_{\alpha n}$ is really small, provided the perturbation $H_{\beta\alpha}$ is small, and masses M_j are large enough in comparison with the mass of the electron. It is essential to mention that in the case when the electron wavefunctions $x_{\alpha}(q,\xi)$ may be chosen real (e.g. in the absence of the external magnetic field), then

$$\left(\frac{\partial}{\partial q_j}\right)_{\alpha\alpha} = \int \chi_{\alpha} \frac{\partial \chi_{\alpha}}{\partial q_j} d\xi = \frac{1}{2} \frac{\partial}{\partial q_j} \int \chi_{\alpha}^2 d\xi = 0$$

Thus we have proved the adiabatic approximation (1.90), (1.91), (1.94) or the Born-Oppenheimer approximation and have found the equation (1.95) describing the breakdown of this approximation. The term

$$A_{\alpha\beta} = H_{\alpha\beta}(1-\delta_{\alpha\beta}) + T_{\alpha\beta}(1-\delta_{\alpha\beta}) - \hbar^2 \sum_{i} \frac{1}{M_i} \left(\frac{\partial}{\partial q_i}\right)_{\alpha\beta} \frac{\partial}{\partial q_i} \tag{1.96}$$

may be considered as an effective perturbation energy describing the deviation from the adiabatic approximation. In the case of the infinite masses of nuclei (apart from the term $H_{\alpha\beta}$) the expressions (1.90), (1.91) and (1.94) with constant $C_{\alpha m}$ are exact solutions of the problem.

Phonons. Phonon-phonon interaction and relaxation

One of the important applications of the Born-Oppenheimer approximation is connected with the description of the crystal lattice. According to the adiabatic approximation we can describe the motion of atoms in crystal separately from the electronic motion. Thus we can introduce the potential energy of the crystal lattice depending only on nuclear coordinates. Having in mind that each atom in crystal has its equilibrium position, we can describe the potential energy of the crystal by the deviations of the nuclear coordinates from their equilibrium positions R^o

$$u_\ell = r_\ell - R_\ell^o$$

Assuming that these deviations are small enough (which is the condition of the stability of the crystal) we can write the Hamiltonian of the crystal (describing nuclear motion in the adiabatic approximation) as follows:

$$H = \frac{1}{2}\sum_\ell M_\ell \dot{u}_\ell^2 + \frac{1}{2}\sum_{\ell_1\ell_2} \Phi_{\ell_1\ell_2} u_{\ell_1} u_{\ell_2} +$$

$$+ \frac{1}{6}\sum_{\ell_1\ell_2\ell_3} \Phi_{\ell_1\ell_2\ell_3} u_{\ell_1} u_{\ell_2} u_{\ell_3} +$$

(Linear in u_ℓ terms do not appear here due to the condition of the equilibrium: $(\partial U_\alpha/\partial q_\ell)_{r_\ell = R_\ell^o} = 0$). As is known, one can always perform the transformation to new coordinates q_k diagonalizing the quadratic terms in the Hamiltonian

$$\mathcal{H} = \frac{1}{2}\sum_k p_k^2 + \omega_k^2 q_k^2 + \sum_{k_1k_2k_3} V_{k_1k_2k_3} q_{k_1} q_{k_2} q_{k_3} + \ldots \quad (1.97)$$

The quasiparticles described by the Hamiltonian

$$\mathcal{H}_0 = \frac{1}{2}\sum_k p_k^2 + \omega_k^2 q_k^2 = \sum_k \left(n_k + \frac{1}{2}\right)\hbar\omega, \quad n_k = a_k^+ a_k \quad (1.98)$$

with energies $h\omega_k$ and with (quasi) momentum $h\vec{k}$ are called phonons. (For more detailed descriptions see any text book on solid state physics). The frequencies of the phonons usually have continuous spectrum and therefore the Hamiltonian (1.97) can be used to describe relaxation processes. The transitions between various phonon states are caused by the unharmonic terms in the Hamiltonian (1.97). Thus, for instance, the term

$$V_{k_1k_2k_3} q_{k_1} q_{k_2} q_{k_3}$$

can cause the transition described by the Golden Rule Formula (1.80). Using this formula and the matrix elements (1.55) one can get

$$w_{n_{k_1}, n_{k_2}, n_{k_3} \to n_{k_1}+1,\, n_{k_2}-1,\, n_{k_3}-1} =$$

$$= \frac{2\pi}{\hbar} \left| V_{k_1 k_2 k_3} \right|^2 \left| q_{k_1\, n_{k_1}, n_{k_1}+1}\, q_{k_2\, n_{k_2}, n_{k_2}-1}\, q_{k_3\, n_{k_3}, n_{k_3}-1} \right|^2 \delta(\hbar\omega_{k_1} - \hbar\omega_{k_2} - \hbar\omega_{k_3})$$

$$= \frac{2\pi}{\hbar} \frac{\left| V_{k_1 k_2 k_3} \right|^2 \hbar^3}{8\, \omega_{k_1} \omega_{k_2} \omega_{k_3}} (n_{k_1}+1)\, n_{k_2}\, n_{k_3}\, \delta(\hbar\omega_{k_1} - \hbar\omega_{k_2} - \hbar\omega_{k_3})$$

The master equation describing the relaxation of the phonons in crystals may be written in the form

$$\dot{\rho}(\{n\}) = -\sum_{\{n'\}} \left[w_{\{n\}\{n'\}}\, \rho(\{n\}) - w_{\{n'\}\{n\}}\, \rho(\{n'\}) \right] \qquad (1.99)$$

where {n} is the manifold of all indices describing phonons (their frequencies, wave-vectors and numbers of phonon branches), and $w_{\{n\}\{n'\}}$ are the transition probabilities per unit time between various phonon states.

II. BASIC PROCESSES AND MODELS

2.1 Adiabatic and non-adiabatic transitions

The Born-Oppenheimer approximation gives the possibility of understanding the essence of chemical transformations and other rate processes, including those occuring in condensed media. According to this approximation, the motion of nuclei, atoms and molecules may be described by the effective potential energy. This potential energy coincides with the electronic energy U_α averaged over the electronic eigenstate α (see (1.91)). For given electronic state α we may describe the motion of nuclei by the Hamiltonian which depends only on nuclear coordinates q

$$\mathcal{H} = T + U_\alpha(q) \tag{2.1}$$

Here T and U_α are kinetic and potential energies of the nuclei. Generally speaking q is a set of nuclear coordinates

$$q = \{q_1, q_2, \ldots, q_N\}$$

In general, the potential energy U_α may have one or several minima which correspond to certain configurations of nuclei, satisfying the conditions

$$\left.\frac{\partial U_\alpha}{\partial q_i}\right|_{q=q^0} = 0 \tag{2.2}$$

These coordinates

$$q^0 = \{q_1^0, q_2^0, \ldots, q_N^0\} \tag{2.3}$$

correspond to stable (or quasi-stable) configuration of nuclei. The force acting on each nucleus is equal to zero in these points

$$F_i = \dot{p}_i = \frac{\partial U}{\partial q_i} = 0$$

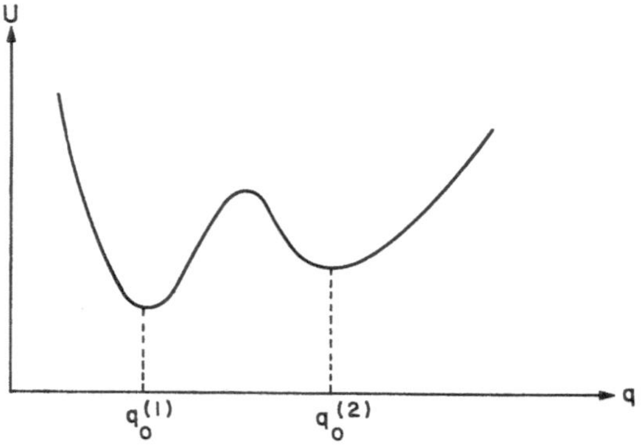

Fig. 1. Potential energy curve with two minima. $q_o^{(1)}$ and $q_o^{(2)}$ - coordinates of the minima.

(Of course, we should remember that (2.2) is a necessary but not a sufficient condition of stability; the sufficient condition of minima is determined by second derivatives of U_α).

Configurations (2.3) describing minima of the potential energy U_α correspond to chemical compounds which is nothing else than stable (or quasi-stable) configuration of nuclei. In these terms we can also understand the meaning of chemical reactions and other rate processes. They correspond to transitions from one stable (or quasi-stable) configuration of nuclei to another one.

Here we have two possibilities. Two configurations, between which we consider the transitions, are at the same electronic state α, i.e. we consider transitions between two stable configurations, two minima on the same potential energy hypersurface. (See Fig. 1).

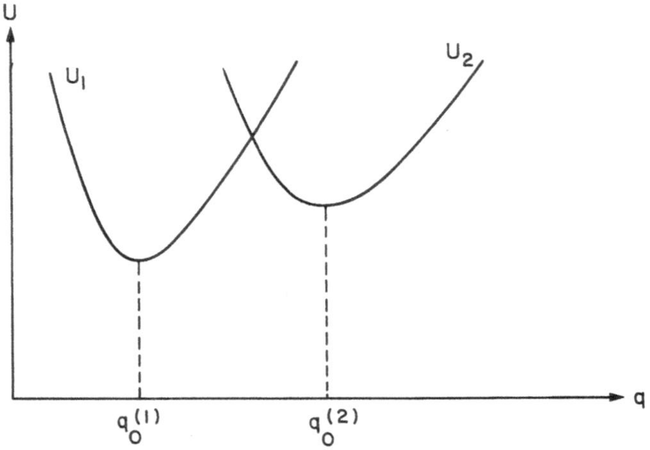

Fig. 2. Two intersecting potential energy curves. Non-adiabatic case.

Such transitions are called adiabatic transitions. Another possibility corresponds to transitions between stable configurations on various electronic states, on various potential energy hypersurfaces U_α and $U_{\alpha'}$, which may be intersecting ones (Fig. 2). Such transitions are called non-adiabatic transitions. This chapter is devoted to various possible mechanisms which may be responsible for transitions between stable configurations of nuclei.

2.2 Adiabatic transitions. Tunneling

One of the first predictions of quantum mechanics was the possibility of transitions between the states divided by the potential barrier, as it is shown in Fig. 3, i.e. "tunneling" through the region of space where the total energy of a particle is less than the potential energy, such a process being impossible in classical mechanics.

The physical importance of this phenomenon has been recognized since the very earliest days of quantum mechanics. Hund [7] discussed the possibility of intermolecular rearrangements via tunneling. Gamov [8] utilized this phenomenon to explain radiative decay. As early as 1932, Wigner [9] discussed tunneling in the context of chemical kinetics.

To understand the phenomenon and for further applications we will start with a very simple system described by the Hamiltonian

$$\mathcal{H} = P^2/2 + U(Q) \tag{2.4}$$

where U(Q) is a one-dimensional potential curve with two minima (see Fig. 3).

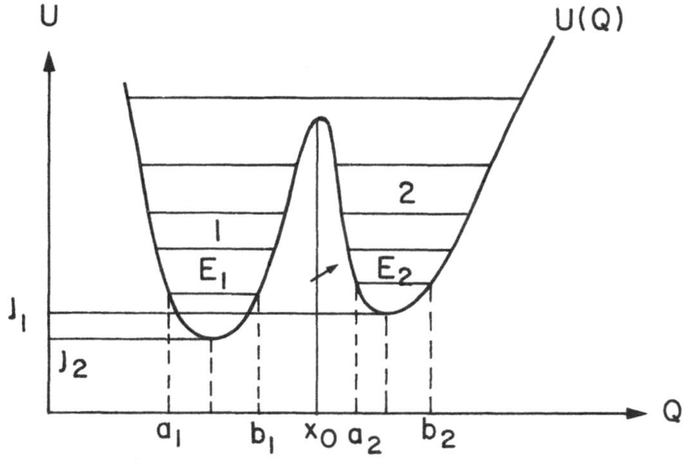

Fig. 3. The adiabatic potential energy curve with two potential wells, E_1 and E_2 - ground states in wells 1 and 2 respectively.

We will consider the transitions between two states with energies E_1 and E_2 close to each other, so that we may neglect the influence of other energy levels. We will assume that the barrier between two wells is large enough, and the tunneling through this barrier may be considered as a small perturbation. Then we can introduce two kinds of states:

(a) Ψ_1 and Ψ_2 with the eigenvalues E_1 and E_2 - which are approximate stationary states (without taking into account the penetration through the barrier) and localized in wells 1 and 2 respectively.

(b) Exact states Ψ_I and Ψ_{II} with the eigenvalues E_I and E_{II}. These states, generally speaking, have no definite localization in the wells 1 and 2 and may be represented as the superposition of the states Ψ_1 and Ψ_2.

Using the results of the theory of two-state systems (see chapter 1.2) and having in mind the fact that for one-dimensional systems the wavefunction could be chosen real, this superposition obtains the form

$$\Psi_I = c_1\Psi_1 + c_2\Psi_2\,, \quad \Psi_{II} = -c_2\Psi_1 + c_1\Psi_2 \tag{2.5}$$

Taking into account only two levels (E_1 and E_2 or E_I and E_{II}) the Hamiltonian (2.4) may be written in two equivalent forms

$$\mathcal{H} = n_1 E_1 + n_2 E_2 + r_+ V_{12} + r_- V_{21} \tag{2.6}$$

or

$$\mathcal{H} = n_I E_I + n_{II} E_{II} \tag{2.7}$$

where the operators $n_{1,2}, r_\pm$ are expressed through the above introduced (in chapter 1.2) operators of effective spin (see (1.41))

$$n_1 = 1/2 - r_3\,, \quad n_2 = 1/2 + r_3\,, \quad r_\pm = r_1 \pm i r_2$$

and V is the effective perturbation energy which is to be calculated now. In good approximation we may write for the wavefunction Ψ_1 and Ψ_2 the Schroedinger equation

$$\Psi_1'' + \frac{2}{\hbar^2}(E_1 - U)\Psi_1 = 0\,, \quad \Psi_2'' + \frac{2}{\hbar^2}(E_2 - U)\Psi_2 = 0 \tag{2.8}$$

These wavefunctions Ψ_1 and Ψ_2 are mainly localized in the wells 1 and 2 respectively and exponentially decay outside the wells. Thus dependence of Ψ_1 and Ψ_2 on Q is determined by spatial dependence of wells 1 and 2 respectively and is almost not affected by the behavior of U far outside the wells. At the same time we may write the exact Schroedinger equations

$$\Psi_I'' + \frac{2}{\hbar^2}(E_I - U)\Psi_I = 0\,, \quad \Psi_{II}'' + \frac{2}{\hbar^2}(E_{II} - U)\Psi_{II} = 0 \tag{2.9}$$

From the equations (2.8) and (2.9) it follows that

$$\int_{x_0}^{\infty} \frac{d}{dQ}\left(\Psi_1'\Psi_I - \Psi_I'\Psi_1\right) dQ = \frac{2}{\hbar^2}\left(E_I - E_1\right)\int_{x_0}^{\infty}\Psi_1\Psi_I\, dQ$$

where x_o is the point in the middle of the barrier (see Fig. 3). From (2.5) and normalization conditions

$$\int_{x_0}^{\infty} |\Psi_1|^2 dQ = 1 \,, \quad \int_{x_0}^{\infty}\Psi_1\Psi_2\, dQ = 0$$

we get

$$c_2\left[\Psi_2'(x_0)\Psi_1(x_0) - \Psi_2(x_0)\Psi_1'(x_0)\right] = \frac{2}{\hbar^2}\left(E_I - E_1\right)c_1 \tag{2.10}$$

On the other hand it follows from the Schroedinger equation (taking $V_{12}=V_{21}=V$)

$$\mathcal{H}\Psi_I = E_I\Psi_I$$

$$\left(n_1 E_1 + n_2 E_2 + 2r_1 V\right)\left(c_1\Psi_1 + c_2\Psi_2\right) = E_I\left(c_1\Psi_1 + c_2\Psi_2\right) \tag{2.11}$$

$$c_1\left(E_1 - E_I\right) + c_2 V = 0$$

Thus from (2.10) and (2.11) we obtain, for the matrix element of the effective perturbation energy, the following expression

$$V = \frac{\hbar^2}{2}\left[\Psi_2'(x_0)\Psi_1(x_0) - \Psi_2(x_0)\Psi_1'(x_0)\right] \tag{2.12}$$

This matrix element may be calculated in the semiclassical approximation:

$$\Psi_1(x_0) = \left(\omega_1/2\pi|P_1|\right)^{1/2}\exp\left(\hbar^{-1}\int_{b_1}^{x_0}|P_1|\,dQ\right)$$

and

$$\Psi_2'(x_0) = -\hbar^{-1}\left(\omega_2|P_2|/2\pi\right)^{1/2}\exp\left(\hbar^{-1}\int_{a_2}^{x_0}|P_2|\,dQ\right),$$

$$P = \left[2(E-U)\right]^{1/2}$$

From these equations and (2.12) it follows that

$$V = \frac{\hbar}{2}(\omega_1\omega_2)^{1/2}\exp\left(\hbar^{-1}\int_{a_2}^{x_0}|P_2|\,dQ + \hbar^{-1}\int_{x_0}^{b_1}|P_1|\,dQ\right)$$

$$\approx \frac{\hbar}{2\pi}(\omega_1\omega_2)^{1/2}\exp\left(-\hbar^{-1}\int_{b_1}^{a_2}|P|\,dQ\right) \tag{2.13}$$

the last expression is valid provided

$$|E_1 - E_2| \ll |U - E_{1,2}|$$

in the region which gives substantial contribution to the integral (2.13). If one of the states 1,2 (or both) is a ground state, we cannot use the semiclassical approach. However, it is still possible to find the correct wave function at the point x_0 without assumption about the semiclassical behavior in the vicinity of the potential wells minima. Instead of it we will make quite reasonable assumptions that in the vicinity of x_0 the semiclassical approximation is good enough and that in the vicinity of the region a_1b_1 (and a_2b_2) the potential well 1 (and 2) may be approximated by the parabola.

The wavefunction, satisfying these assumptions, may be represented (far enough from the minimum of the potential wells) as follows

$$\Psi = \left(\frac{\pi}{e}\right)^{1/4}\left(\frac{\omega_1}{2\pi}\right)^{1/2}\left[2\left(U - J_1 - \tfrac{1}{2}\hbar\omega_1\right)\right]^{-1/4}\exp\left[-\hbar^{-1}\int_{(\hbar/\omega_1)^{1/2}}^{Q}\left[2\left(U_1 - J_1 - \tfrac{1}{2}\hbar\omega_1\right)\right]^{1/2}dQ\right]$$

where $J_1 = E_1 - \hbar\omega_1/2$; $J_2 = E_2 - \hbar\omega_2/2$ are the meanings of the potential energies U_1 and U_2 at the points of their minima.

In the region where

$$U_1 - J_1 \approx \frac{1}{2}\omega_1 Q^2 \quad \text{and} \quad \omega_1^2 Q^2/2\hbar\omega_1 >> 1$$

(if such a region exists)

$$\Psi_1 \approx (\omega_1/\pi\hbar)^{1/4}\exp\left(-\omega_1 Q^2/2\hbar\right)$$

which coincides with the exact ground state for the harmonic oscillator and proves the correct choice of the wavefunction.

Thus, if both states E_1 and E_2 are ground states we should write, instead of (2.13)

$$V_{12} = V_{21} = V = \frac{\hbar}{2}(\omega_1\omega_2/\pi e)^{1/2} \exp\left(-\hbar^{-1}\int_{b_1}^{a_2} |P|\, dQ\right) \tag{2.14}$$

Now, having the Hamiltonian (2.6) and the perturbation energy matrix elements (2.13), (2.14) we are able to solve simple problems relating to the tunneling.

(1) Quantum beats

Let us suppose that initially the particle was in one of the wells 1 and 2 (see Fig. 3) i.e., in one of the states 1 or 2. The question arises about the time evolution of the system. Such time evolution may be described by the 2x2-density matrix

$$\sigma = \begin{pmatrix} \sigma_{11} & \sigma_{12} \\ \sigma_{21} & \sigma_{22} \end{pmatrix}$$

From the von Neumann equation

$$i\hbar \frac{\partial \sigma}{\partial t} = [H, \sigma]$$

and the Hamiltonian (2.6) we get

$$\begin{aligned} \dot{\sigma}_{11} &= -\dot{\sigma}_{22} = -i\hbar^{-1} V (\sigma_{21} - \sigma_{12}) \\ \dot{\sigma}_{12} &= \dot{\sigma}_{21}^{*} = -i\omega_{12}\sigma_{12} - i\hbar^{-1} V (\sigma_{22} - \sigma_{11}) \end{aligned} \tag{2.15}$$

where $\omega_{12} = (E_1 - E_2)/\hbar$.

Performing simple manipulations we get for the population difference

$$n = \sigma_{11} - \sigma_{22}$$

the equation

$$\dddot{n} + \Omega^2 \dot{n} = 0 \tag{2.16}$$

where (see also (1.49))

$$\Omega = \hbar^{-1}\left(4V^2 + \hbar^2\omega_{12}^2\right)^{1/2} \tag{2.17}$$

The equation (2.16) has a general solution

$$n = A\cos(\Omega t) + B\sin(\Omega t) + C$$

If in the initial time $t = t_o$ the density matrix was diagonal

$$\sigma_{12}(t_o) = \sigma_{21}(t_o) = 0$$

then

$$n(t) = n(t_o)\left[4V^2\hbar^{-2}\cos(\Omega(t-t_o)) + \omega_{12}^2\right]\Omega^{-2} \tag{2.18}$$

Thus the population difference of two wells 1 and 2 varies harmonically with the frequency Ω (2.18) which depends on the matrix element V and ω_{12}. Of course, this result could also be obtained from the relations (1.45), (1.51) valid for an arbitrary two-state system.

In the case of symmetrical wells, when

$$\omega_{12} = 0$$

the frequency of the so-called quantum beats is equal

$$\Omega_o = 2V/\hbar \tag{2.19}$$

and the amplitude of beats varies from $n(t_o)$ to $-n(t_o)$ (if at $t = t_o$, $n = 1$ then it varies from 1 to -1). It means that if initially the particle was at well 1, then at time

$$t - t_o = \pi/\Omega_o$$

the particle would be found in the well 2 with probability equal unity. Then in the time interval

$$\Delta t = \pi/\Omega_o = \pi\hbar/2V \tag{2.20}$$

the particle would come back to the first well and so on.

On the other hand when

$$|\omega_{12}| \gg 2V/\hbar \tag{2.21}$$

the quantum beats are suppressed and the amplitude of the variation of the population difference n is quite small

$$n(t) \approx n(t_o)\left[1 + (2V/\hbar\omega_{12})^2\left[\cos(\Omega(t-t_o)) - 1\right]\right] \tag{2.22}$$

and

$$|n - n(t_0)| \simeq (2V/\hbar\omega_{12})^2$$

It means that all the time the probability of finding the particle in the well 1 is close to unity and the probability of finding the particle in the well 2 is small, and has the order of magnitude $V^2/\hbar^2\omega_{12}^2$.

(2) Escaping of particle from the potential well

Now let us consider the potential curve with quite asymmetric wells (see Fig. 4). We should keep in mind the situation when the dimension of the well 2 tends to infinity and respectively, the energy differences

$$\Delta E = \hbar\omega_2 \rightarrow 0$$

so that energy levels of the well 2 form quasicontinuous spectrum (it means that condition (1.82) should be satisfied). It should be stressed that consideration of this kind of potential curve is an essential deviation from our major topic connected with rate processes in condensed medium.

In condensed medium particles cannot perform infinite movement as it is implied from the potential energy curve in the region 2. However, it seems instructive to consider this kind of problem for the sake of comparison.

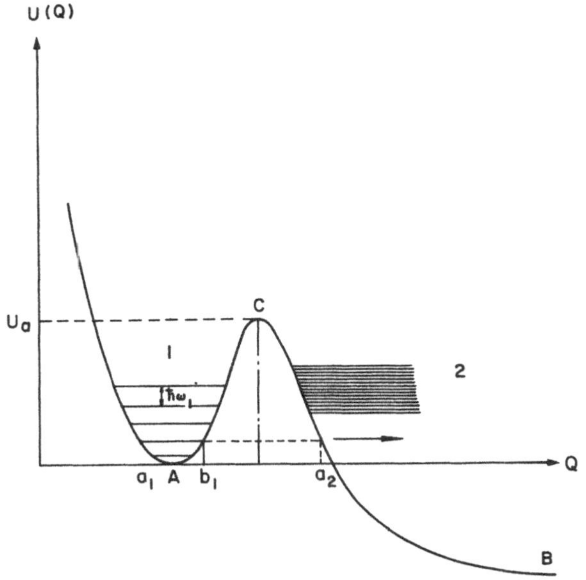

Fig. 4. The potential energy curve which may be appropriate for modelling rate processes in gaseous media. In the region 2 the energy spectrum is quasicontinuous and corresponds to infinite motion.

Now, the level 1 in the well 1 is degenerated with the quasicontinuum spectrum of energies and we can use the Golden Rule Formula (1.80). The rate of escape from the well 1 is equal

$$w = \frac{2\pi}{\hbar} \int |V|^2 \delta(E_1 - E_2)\, \rho_2(E_2)\, dE_2 = \frac{2\pi}{\hbar} |V|^2 \rho_2(E_2) \quad (2.23)$$

To find the density of states E_2 it is enough to recall that in the semiclassical case the energy differences are equal

$$\Delta E_2 = \Delta n_2 \hbar \omega_2$$

Therefore the number of states in the energy interval ΔE_2 is equal

$$\Delta n_2 = \Delta E_2 / \hbar \omega_2$$

i.e.

$$\rho_2(E_2) = 1/\hbar \omega_2(E_2) \quad (2.24)$$

Substituting (2.13), (2.24) for (2.23) we obtain

$$w = \frac{\omega_1}{2\pi} \exp\left(-2\hbar^{-1} \int_{b_1}^{a_2} |P|\, dQ\right) = \frac{\omega_1}{2\pi} D \quad (2.25)$$

where

$$D = \exp\left(-2\hbar^{-1} \int_{b_1}^{a_2} |P|\, dQ\right) \quad (2.26)$$

is the so-called transmission coefficient, having the meaning of the probability of transmission through the barrier. Thus, formula (2.25) has a very simple interpretation: the transition probability per unit time is equal to the frequency of encounters of particle with barrier $\omega_1/2\pi$ times the probability of transmission D through the barrier. Of course, such a simple explanation is connected with the semiclassical character of the problem. On the other hand, from (2.23), (2.25) and (2.24) we can express the matrix element of perturbation through the transmission coefficient

$$|V|^2 = \hbar^2 \omega_1 \omega_2 D (2\pi)^{-2} \quad (2.27)$$

At finite temperature of the medium the particle may escape from the potential well 1 (Fig. 4) overcoming the potential barrier by the thermal fluctuation.

Such mechanism is a basis of a very popular method of calculating the reaction rate constants in chemistry. The method is called the transition state method, or the activated complex method, or the theory of absolute reaction rate constants and is associated with the names of Eyring, Polany and Wigner. According to the transition-state-method one considers the particles near A to be in perfect temperature equilibrium with those near B so that we have thermal equilibrium also at C (Fig. 4); and one calculated the number of particles which is in unit time pass the transition point C from left to right. This number is, of course, equal to that passing from the right to the left and is given by

$$k_{12} = \frac{2\pi}{\hbar}\Big(\sum_s e^{-E_{1s}/k_BT}\Big)^{-1} \sum_{l,n} e^{-E_{1l}/k_BT} |V_{ln}|^2 \delta(E_{1l} - E_{2n}) =$$

$$= \frac{2\pi}{\hbar} \iint (dE_1/\hbar\omega_1)(dE_2/\hbar\omega_2)\left[\hbar^2\omega_1\omega_2/(2\pi)^2\right] D\, \delta(E_1-E_2)\, e^{-E_1/k_BT} \Big/ \Sigma_1 \tag{2.28}$$

where Σ_1 is the statistical sum of the well 1

$$\Sigma_1 = \sum_n{}' \exp(-n\hbar\omega_1/k_BT) \tag{2.29}$$

We will now consider the case when the major role plays not the tunneling but classical processes, i.e., overcoming of the barrier induced by the thermal fluctuations. It means that we consider the temperatures high enough in comparison with $h\omega_1/k_B$

$$k_BT \gg \hbar\omega_1 \tag{2.30}$$

On the other hand we consider the temperature to be low enough in comparison with the height of the barrier U_a (see Fig. 4).

$$k_BT \ll U_a \tag{2.31}$$

Then it is easy to see that

$$\Sigma_1 = \sum_n{}' \exp(-n\hbar\omega_1/k_BT) = \left[1 - e^{-\hbar\omega_1/k_BT}\right] \simeq k_BT/\hbar\omega_1 \tag{2.32}$$

At the temperatures (2.30) we may take into account only classical transitions, i.e., to approximate D by the step-function

$$D = \begin{cases} 0, & E_1 < U_a \\ 1, & E_1 > U_a \end{cases} \tag{2.33}$$

Then the expression (2.28) obtains the form

$$k_{12} = (2\pi)^{-1}(\omega_1/k_B T)\int_{U_a}^{\infty} e^{-E_1/k_B T} dE_1 = \frac{\omega_1}{2\pi} e^{-U_a/k_B T} \tag{2.34}$$

This formula is a basis of the transition state method.

The assumptions under which formula (2.34) has been derived are not very clear and there is doubt whether they may be justified at all. There exists extensive literature devoted to the transition state method.

Here I want to mention some basic works and latest papers. The book of S. Glasstone, U.J. Laidler and H. Eyring [10] is the basic monography devoted to the subject. Critical analysis of the assumptions of the transition state method was performed in the pioneering work of Kramers [11].A general review devoted to the subject and many other stochastic problems,was done by Chandrasekhar [12]. Among recent works devoted to the subject are the papers [13-19].

2.3 Non-Adiabatic transitions. Landau-Zener transition

According to the definition, the non-adiabatic transitions are those between the stable configurations on various electronic states (see Fig. 2). The theory of elementary acts of such transition was given by Landau and Zener and is called:

The Landau-Zener transition

Let U_1 and U_2 be two electronic terms, i.e., according to Chapter 1.4 they are the energy eigenvalues of the approximate electronic Hamiltonian $H_o(q,\xi)$ (1.90).

These eigenvalues are equal

$$U_1 = E_1(q), \quad U_2 = E_2(q)$$

Let V be the perturbation energy. For example, if E_1 and E_2 correspond to definite spin states, the perturbation energy may be the spin-orbit coupling. We would consider now V to be the main cause for transition between the states 1 and 2. (Another possibility is the transition caused by the terms in (1.95) proportional to the derivatives $\partial/\partial q$, i.e. connected with the break-down of the Born-Oppenheimer approximation).

To each potential curve U_1 and U_2 correspond energy levels (e.g. vibrational states) E_{1n} and $E_{2n'}$. To find the transition probability between two states (with equal energy)

$$E_{1n} = E_{2n'}$$

we should know the perturbation energy matrix element. Such matrix element may be found in semiclassical approximation.

We would assume that $E = E_{1n} = E_{2n}$, is larger than the energy E_o, corresponding to the point of intersection of these two curves U_1 and U_2 (see Fig. 5).

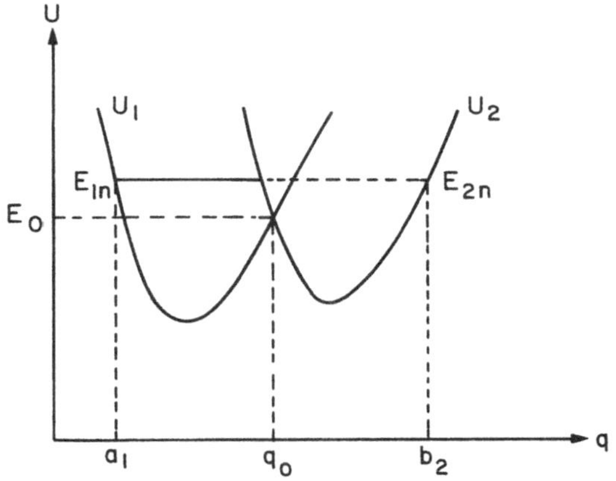

Fig. 5. Two intersecting potential energy waves U_1 and U_2; E_{1n} and E_{2n} are the energy levels corresponding to these potentials. E_o - the energy level corresponding to the intersection point q_o

Transition with energies $E < E_o$ correspond to penetration through classically inaccessable regions and their probabilities are much smaller than those with energies $E > E_o$. In the semiclassical approximation (see Fig. 5), the eigenfunctions of the states 1 and 2, in the regions $(a_1 b_1)$ and $(a_2 b_2)$ respectively, are equal

$$\psi_1 = (2\omega_1/\pi v_1)^{1/2} \cos\left(\hbar^{-1}\int_{a_1}^{q} p_1\, dq - \pi/4\right)$$
$$\psi_2 = (2\omega_2/\pi v_2)^{1/2} \cos\left(\hbar^{-1}\int_{a_2}^{q} p_2\, dq - \pi/4\right) \qquad (2.35)$$

where

$$E_1 = (n_1 + 1/2)\hbar\omega_1, \quad E_2 = (n_2 + 1/2)\hbar\omega_2$$

$$\omega = 2\pi/T, \quad T = 2\int_a^b v^{-1} dq, \quad v = p/M$$

$$p = (2M(E-U))^{1/2}, \quad 2\int_a^b p\, dq = (n + 1/2)\hbar$$

Thus the matrix element V_{12} obtains the form

$$V_{12} = \frac{2}{\pi}\int dq \left(\frac{\omega_1 \omega_2}{v_1 v_2}\right)^{1/2} V(q) \cos\left(\hbar^{-1}\int_{a_1}^{q} p_1 dq - \frac{\pi}{4}\right)\cos\left(\hbar^{-1}\int_{a_2}^{q} p_2 dq - \frac{\pi}{4}\right) \quad (2.36)$$

The product of cosines may be presented as

$$2\cos\left(\hbar^{-1}\int_{a_1}^{q} p_1 dq - \frac{\pi}{4}\right)\cos\left(\hbar^{-1}\int_{a_2}^{q} p_2 dq - \frac{\pi}{4}\right) =$$

$$= \cos\left(\hbar^{-1}\int_{a_1}^{q} p_1 dq - \hbar^{-1}\int_{a_2}^{q} p_2 dq\right) + \cos\left(\hbar^{-1}\int_{a_1}^{q} p_1 dq + \hbar^{-1}\int_{a_2}^{q} p_2 dq\right)$$

Now we would utilize the property of the semiclassical approximation according to which the integral $\hbar^{-1}\int_{a_1}^{q} p_1 dq$ is large ($n_1, n_2 \gg 1$). This means that cosines are high oscillating functions in all points except the point where the derivative of the argument is equal zero. Such a point exists in the first cosine and the vicinity of this point gives the main contribution to the integral. This point is determined by the expressions

$$S = \hbar^{-1}\int_{a_1}^{q} p_1 dq - \hbar^{-1}\int_{a_2}^{q} p_2 dq \,, \quad S' = \hbar^{-1}(p_1 - p_2) = 0$$

$$p_1(q_0) = p_2(q_0)\,, \quad E_1 - U_1 = E_2 - U_2 \quad (2.37)$$

Taking into account that $E_1 = E_2$, it is easy to see that point q_o giving the main contribution to the integral (2.36) is just the point of the intersection of the potential curves U_1 and U_2

$$U_1(q_0) = U_2(q_0) \quad (2.38)$$

The next assumption of the Landau-Zener approximation is that we can neglect the variation of the perturbation energy in the vicinity of the point $q = q_o$.

Expanding the argument of the cosine up to terms ξ^2, where $\xi = q - q_o$ the matrix element (2.36) obtains the form

$$V_{12} = \frac{(\omega_1 \omega_2)^{1/2} V(q_0)}{\pi \; v(q_0)} \int_{-\infty}^{\infty} \cos\left[\hbar^{-1} S_0 + (2\hbar)^{-1}\left(\frac{\partial p_1}{\partial q} - \frac{\partial p_2}{\partial q}\right)_{q=q_0} \xi^2\right] d\xi$$

Then, making use of the relations

$$p_1^2/2M + U_1 = p_2^2/2M + U_2 , \quad v_1 \partial p_1/\partial q - v_2 \partial p_2/\partial q = F_1 - F_2$$

$$F = -\partial U/\partial q , \quad S = S_0 + [(F_1 - F_2)/2v]\xi^2$$

$$\int_{-\infty}^{\infty} \cos(\alpha + \beta\xi^2)\, d\xi = (\pi/\beta)^{1/2} \cos(\alpha + \pi/4)$$

we get the following expression for the matrix element V_{12}

$$V_{12} = V\left[2\hbar\omega_1\omega_2/\pi v\,|F_1 - F_2|\right]^{1/2} \cos\left(\hbar^{-1} S_0 + \pi/4\right) \tag{2.39}$$

The quantity $S/\hbar$ is very large and cosine varies rapidly with the change of energy. This means that by averaging $\cos^2$ $(S/h + \pi/4)$ over small intervals of energies it is possible to replace $\cos^2$ by its average value 1/2.

Thus, finally, we obtain the formula for the matrix element in the Landau-Zener approximation

$$V_{12}^2 = V^2\hbar\omega_1\omega_2/\pi v\,|F_1 - F_2| \tag{2.40}$$

where $\tilde{v}$ is taken at the point of the interseciton of the curves U_1 and U_2.

Now, if the curve U_2 corresponds to infinite motion of the particle (see Fig. 6)

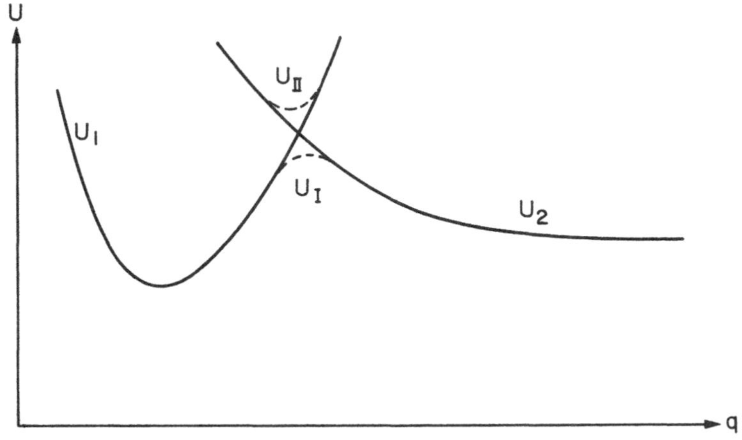

Fig. 6. Two intersecting potential energy curves. The curve U_2 corresponds to infinite motion. U_I and U_{II} are adiabatic potential energy curves.

the spectrum of its energies is quasi-continuous, and we can calculate the transition probability per unit time (using relations (1.80), (2.24))

$$w_{12} = \frac{2\pi}{\hbar} V_{12}^2 \, \rho(E_2) = 2\omega_1 V^2/\hbar v \, |F_1 - F_2| \tag{2.41}$$

The applicability of the perturbation theory implies that this transition probability should be small in comparison to the frequency of unperturbed motion in the well 1

$$w_{12}/\omega_1 = 2V^2/\hbar v \, |F_1 - F_2| \ll 1 \tag{2.42}$$

The smallness of this parameter also determines the applicability of the nonadiabatic approximation.

When this parameter is large we cannot use as a first approximation intersecting potential energy curves and we should take into account change of the energy curves due to the interaction term V. It is easy to show that in this case instead of using curves U_1 and U_2 we should use the following adiabatic non-intersecting potential energy curves

$$U_{I,II} = \frac{1}{2}(U_1 + U_2) \pm \frac{1}{2}\left[(U_1 - U_2)^2 + 4V^2\right]^{1/2} \tag{2.43}$$

(see Fig. 6)

In the case (suitable for condensed media) when both potential wells correspond to finite motion, we cannot introduce the transition probability per unit time. In this case the energy spectrum is discrete. Without taking into account an interaction with condensed medium the motion is reversible and relaxation processes do not take place. In the next chapters we will see how the medium modifies the situation.

2.4 Electron transfer

One of the important elementary processes going on in condensed media is the electron transfer [20, 21, 22]. Here we want to present the quantum-theoretical descriptions of this process [21]. Let ε_1 and ε_2 be the electronic states, each of which correspond to different localizations ξ_1 and ξ_2 of the electron (Fig. 7). Then transition from the state ε_1 to the state ε_2 would be accompanied by the electron transfer from the point ξ_1 (or to be more exact, from the vicinity of this point) to the point ξ_2. In the adiabatic approximation the electronic energies may be considered as functions of nuclear configurations (see Fig. 7). This also means that the electron transfer essentially depends on the nuclear configuration and is accompanied by the change of this configuration. Of course, such change may be connected with transfer of nuclei in the molecule or with change of the configuration in the surrounding media, e.g. in the solution or in the huge biomolecule.

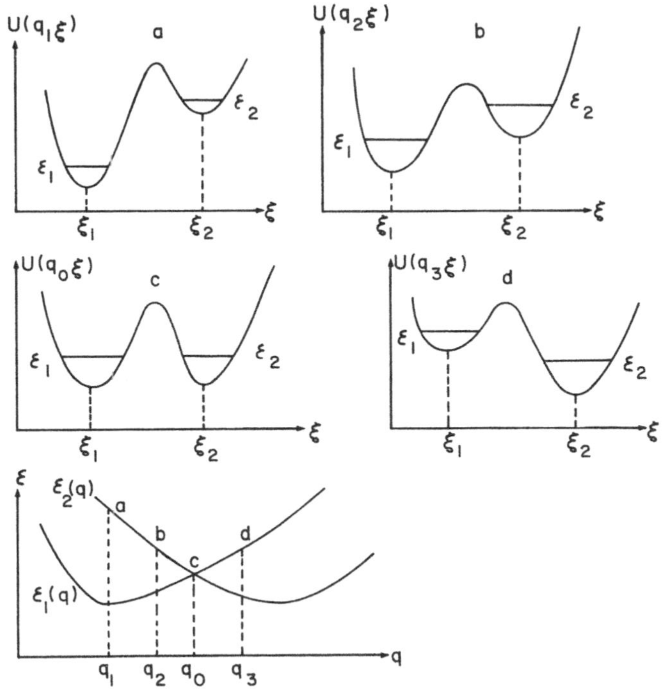

Fig. 7. Potential energies of electron $U(q,\xi)$ and energy levels ε_1 and ε_2 as function of nuclear coordinates. Curves a,b,c,d correspond to various meanings of the nuclear coordinate q. The lower curve describes dependence of the electronic energies ε_1 and ε_2 on nuclear coordinates q.

Why is it necessary that the change of the nuclear configuration should occur? Let us look at the Fig. 7a, which corresponds to the configuration of nuclei $q = q_1$. In this configuration there is a large gap between energies ε_1 and ε_2. This means that the process of tunneling between these states has very low probability. But if the configuration of nuclei changes from q_1 to q_0, then the energy gap is equal to zero and the tunneling will have its maximum value. Thus nuclear motion stimulates the electron transfer.

In the condensed media the initial and final states will be characterized by the electronic energy hypersurfaces (with an infinite number of degrees of freedom)

$$U_1 = \mathcal{E}_1(q_1, q_2, \dots q_N) \;,\; U_2 = \mathcal{E}_2(q_1, q_2, \dots q_N) \tag{2.44}$$

Transitions between these hypersurfaces imply the electron transfer.

Now the Hamiltonian of the system (including electronic variables and taking into account only two electronic states) may be written in the Born-Oppenheimer approximation as

$$\mathcal{H}=\sum_{k=1}^{N} p_k^2/2M_k+n_1U_1(q_1,q_2,\ldots q_N)+n_2U_2(q_1,q_2,\ldots,q_N)+r_+V_{12}+r_-V_{21} \quad (2.45)$$

where

$$n_1=1/2-r_3\ ,\quad n_2=1/2+r_3\ ,\quad r_\pm=r_1\pm ir_2 \qquad \text{(see (1.4.))}$$

In order to carry out an analytical description, we will utilize widely accepted model in which the energies U_1 and U_2 are the sums of energies of harmonic oscillators with identical frequencies but different equilibrium positions

$$U_1=J_1+\frac{1}{2}\sum_{k=1}^{N}\omega_k^2(q_k-q_{1k}^0)^2,\quad U_2=J_2+\frac{1}{2}\sum_{k=1}^{N}\omega_k^2(q_k-q_{2k}^0)^2 \quad (2.46)$$

Thus the Hamiltonian (2.45) in this model may be rewritten (using units in which the kinetic energy is equal to $\Sigma\ p_k^2/2$).

$$\mathcal{H}=\frac{1}{2}\sum_k(p_k^2+\omega_k^2q_k^2)+n_1E_1+n_2E_2$$
$$-n_1\sum_k\omega_k^2q_{1k}^0q_k-n_2\sum_k\omega_k^2q_{2k}^0q_k+r_+V_{12}+r_-V_{21} \quad (2.47)$$

where

$$J_1=E_1-\frac{1}{2}\sum_k\omega_k^2q_{1k}^{0\,2},\quad J_2=E_2-\frac{1}{2}\sum_k\omega_k^2q_{2k}^{0\,2} \quad (2.48)$$

For further applications we will perform the well known transformation eliminating terms of $-n_i\Sigma\omega_k^2q_{ik}^0q_k$ type. Such a unitary transformation is realized by the operator

$$U=\prod_k\exp\left[-\frac{i}{\hbar}(q_{1k}^0n_1+q_{2k}^0n_2)p_k\right] \quad (2.49)$$

As can be easily checked, the transformed Hamiltonian, for which we shall preserve its former designation, takes the form

$$U^{-1}\mathcal{H}U \to \mathcal{H} = n_1 J_1 + n_2 J_2 + \frac{1}{2}\sum_k \left(p_k^2 + \omega_k^2 q_k^2\right) + r_+ \Pi_1 V_{12} \Pi_2^+ + r_- \Pi_2 V_{21} \Pi_1^+ = \mathcal{H}_0 + V \tag{2.50}$$

where

$$\Pi_j = \prod_k \exp\left(\frac{i}{\hbar} q^0_{jk} p_k\right) = \prod_k \exp\left[-\lambda_{jk}\left(a_k^+ - a_k\right)\right] \tag{2.51}$$

and

$$\lambda_{jk} = \left(\omega_k / 2\hbar\right)^{1/2} q^0_{jk} ,$$

a_k, a_k^+ are the phonon annihilation and creation operators introduced in chapter 1.2 (see (1.56), (1.57)). Further on we will consider the simple case when

$$V_{12} = V_{21} = V \tag{2.52}$$

and does not depend on q. In this case the interaction energy has the form

$$\hat{V} = r_+ V \Pi^+ + r_- V \Pi \tag{2.53}$$

where

$$\Pi = \prod_k \exp\left[-\lambda_k\left(a_k^+ - a_k\right)\right], \quad \lambda_k = \left(\omega_k / 2\hbar\right)^{1/2}\left(q^0_{1k} - q^0_{2k}\right) \tag{2.54}$$

The Hamiltonian (2.50) may serve as a starting point for the calculation of the transition probabilities and for consideration of the time behavior of the system.

Now it is easy to see that the problem of electron transfer is reduced to that of the transition from state 1, characterized by electronic number 1 (energy J_1) and quantum number of phonons α to the state 2 with another quantum number α' of phonons. The transition probability averaged over the initial state $\rho_{1\alpha 1\alpha}$ has the form (see (1.80)).

$$w = \frac{2\pi}{\hbar}\sum_{\alpha,\alpha'} \rho_{1\alpha,1\alpha} V_{1\alpha,2\alpha'} V_{2\alpha',1\alpha}\, \delta\left(J_1 - J_2 + F_\alpha - F_{\alpha'}\right)$$

where F_α and $F_{\alpha'}$ are the energy levels of the phonon system. For further applications it is worthwhile to make some simple transformations and to present the transition probability in a different form. For this we will use the identity

$$\delta(J_1-J_2+F_\alpha-F_{\alpha'}) = (2\pi\hbar)^{-1}\int_{-\infty}^{\infty} dt\, \exp\left[i(J_1-J_2+F_\alpha-F_{\alpha'})t/\hbar\right]$$

Now we may write

$$w_{12} = \hbar^{-2}\int_{-\infty}^{\infty}\sum \rho_{1\alpha 1\alpha} V_{1\alpha 2\alpha'} e^{i(F_\alpha - F_{\alpha'})t/\hbar} V_{2\alpha' 1\alpha} e^{i\omega_{12}t}\, dt =$$
$$= \hbar^{-2}\int_{-\infty}^{\infty} e^{i\omega_{12}t}\langle V_{12}(t)V_{21}(0)\rangle\, dt \tag{2.55}$$

where $\omega_{12} = (J_1-J_2)/\hbar$; $\langle\ldots\rangle$ means averaging over the initial phonon states and the time dependence $V_{12}(t)$ is determined here by the Hamiltonian

$$F = \frac{1}{2}\sum_k (p_k^2 + \omega_k^2 q_k^2)$$

Substituting (2.53) (2.54) for (2.55) we get

$$w_{12} = V^2\hbar^{-2}\int_{-\infty}^{\infty} e^{i\omega_{12}\tau}\langle \Pi^+(\tau)\Pi(0)\rangle\, d\tau =$$
$$= V^2\hbar^{-2}\int_{-\infty}^{\infty} e^{i\omega_{12}\tau}\langle \prod_k \exp\left[\eta_k(a_k^+(\tau)-a_k(\tau))\right]\exp\left[-\eta_k(a_k^+-a_k)\right]\rangle d\tau \tag{2.56}$$

To simplify this expression we would utilize the fact that each of η_k is infinitesimally small, provided that there are no singled out vibrations (like a local vibration). It follows from the relations (2.48) that the sum $\Sigma\omega_k^2 q_{jk}^{o2}$ should be finite, which is possible if $q_{jk}^{o} \propto N^{-\frac{1}{2}}$, and the same refers to η_k.

In this case (in the limit $N \to \infty$) the integrand (2.56) may be transformed as follows

$$\prod_k \langle \left[\left(1+\eta_k(a_k^+(\tau)-a_k(\tau)\right)+\frac{1}{2}\eta_k^2\left(a_k^+(\tau)-a_k(\tau)\right)^2\right]\left[1-\eta_k(a_k^+-a_k)\right.$$
$$\left.+\frac{1}{2}\eta_k^2(a_k^+-a_k)^2\right]\rangle = \exp\left[-\sum_k\left(\eta_k^2(2\bar{n}_k+1)-\eta_k^2\left(\bar{n}_k e^{i\omega_k\tau}+(\bar{n}_k+1)e^{-i\omega_k\tau}\right)\right)\right]$$

Here we have assumed that

$$\langle a_k^+ a_k^+\rangle = \langle a_k a_k\rangle = 0,\; a_k^+(t) = a_k e^{i\omega_k t},\; a_k(t) = a_k e^{-i\omega_k t}$$

It may be shown that the relation

$$\langle \Pi^{+}(t)\Pi(0)\rangle = \exp\Big[-\sum_k \eta_k^2(2\bar{n}_k+1)+\sum_k \eta_k^2\big(\bar{n}_k e^{i\omega_k\tau}+(\bar{n}_k+1)e^{-i\omega_k\tau}\big)\Big] \tag{2.57}$$

is also valid in the case when the averaging is performed over the thermodynamic equilibrium. In this case there is no need for the assumption of the smallness of η_k.

Thus the transition probability between two electronic states obtains the form

$$w_{12} = \frac{V^2}{\hbar^2} e^{-\sum_k \eta_k^2(2\bar{n}_k+1)} \int_{-\infty}^{\infty} d\tau\, e^{i\omega_{12}\tau} \exp\Big[\sum_k \eta_k^2\big(\bar{n}_k e^{i\omega_k\tau}+(\bar{n}_k+1)e^{-i\omega_k\tau}\big)\Big] \tag{2.58}$$

This expression takes into account various multiphonon transitions between electronic states 1 and 2. In the case when the exponent

$$\sum_k \eta_k^2\big[\bar{n}_k e^{i\omega_k\tau}+(\bar{n}_k+1)e^{-i\omega_k\tau}\big]$$

is small, the main contribution gives zero and one-phonon transitions (if they are allowed by the energy conservation laws):

$$w_{12} = \frac{V^2}{\hbar^2}\int_{-\infty}^{\infty} d\tau\Big[e^{i\omega_{12}\tau}+\sum_k e^{i\omega_{12}\tau}\eta_k^2\big[\bar{n}_k e^{i\omega_k\tau}+(\bar{n}_k+1)e^{-i\omega_k\tau}\big]\Big]$$

$$= \frac{2\pi}{\hbar}V^2\delta(J_1-J_2)+\frac{2\pi}{\hbar}V^2\sum_k\big[\bar{n}_k\delta(J_1-J_2+\hbar\omega_k)+ \tag{2.59}$$

$$+(\bar{n}_k+1)\delta(J_1-J_2-\hbar\omega_k)\big]$$

It should be stressed that the first term describing the transition between two discrete electronic levels, without the phonons, has no sense, since the transition probability per unit time in this case does not exist. The second term describes either transition with the absorption of one phonon (if $J_1 < J_2$, the first term $\propto \bar{n}_k$) or with spontaneous and stimulated emission of phonons (if $J_1 > J_2$, the term $\bar{n}_k + 1$). When the exponent $\Sigma\eta_k^2[\bar{n}_k e^{i\omega_k\tau} + (\bar{n}_k+1)e^{i\omega_k\tau}]$ is not small, we should take into account all multiphonon transitions. In this case the calculation of the expression becomes more complicated and various approximations may be used. Below we will review the possible methods of calculating the transition probability (2.58).

2.5 Energy transfer

The energy transfer between atoms (molecules) embedded in condensed media or energy transfer between different sub-units of a macro-molecular framework is of vital importance to diverse fields of research such as sensitized luminescence and photosynthesis. No doubt the understanding and theory of the energy transfer has biological importance. Starting from pioneering works of Förster [23] and Dexter [24], a lot of theoretical and experimental works devoted to this problem have been done. Reviews of the problem and the bibliography are contained in the papers [25-30]. One of the simplest problems connected with electronic energy transfer is that of a dimer. The dimer is a two-molecular chain and therefore must have a relationship to the problem of energy transfer in an infinite molecular chain, i.e to excitation propagation problems in molecular crystal. On the other hand, the theory of the energy transfer between two sub-units of a macro-molecular framework may have direct relation to biological problems.

In this section we want to present the theoretical framework for description of the energy transfer in condensed media. The Hamiltonian describing energy transfer between two atoms (molecules) embedded in a condensed medium was discussed by Soules and Duke [31], Rackovsky and Silbey [32] and Abram and Silbey [33]. We will deal with the simplified model, taking into account only two levels of each molecule. We will neglect transitions between energy levels of each molecule and take into account only the energy transfer process between these molecules caused by the intermolecular interaction V (see Fig. 8). The two-molecule system in this

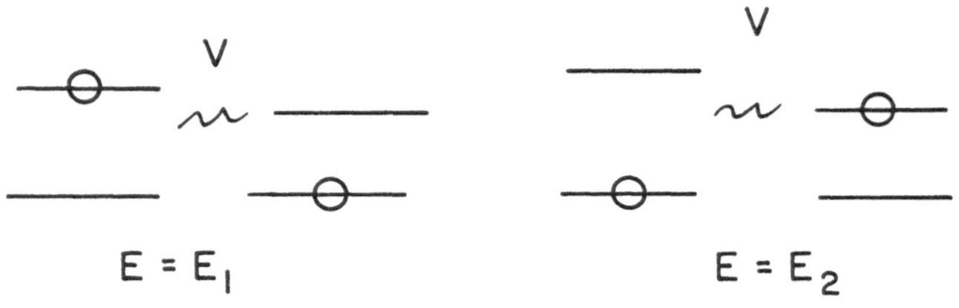

Fig. 8. Energy levels of two molecules coupled by the interaction energy V.

case may be in two possible states.

In state 1, the first molecule is in the excited state and the second molecule is in the ground state. In state 2, the second molecule is in the excited and the first molecule is in the ground state. Respectively, the energy of the first state is E_1 and the second state is E_2. Thus we have come to the situation with which we dealt in Chapter 1.2, i.e. with a two-state system. The Hamiltonian of such a system may be presented in the form (1.40).

$$E = E_1 n_1 + E_2 n_2 + r_+ V_{12} + r_- V_{21} \tag{2.60}$$

In particular, the interaction energy V may coincide with that of dipole-dipole interaction between the molecules.

We assume that the molecules interact with vibrational excitations of media (phonons) through linear electron-phonon coupling. Then, the Hamiltonian of the whole system, including condensed medium, may be written in the form

$$\begin{aligned}\mathcal{H} = {} & n_1 E_1 + n_2 E_2 + r_+ V_{12} + r_- V_{21} + \sum_{\vec{q}}' \hbar\omega_{\vec{q}} \left(b^+_{\vec{q}} b_{\vec{q}} + \frac{1}{2}\right) \\ & + \sum_{\vec{q}}' \hbar\omega_{\vec{q}} (b_{\vec{q}} + b_{-\vec{q}}) \Big\{ n_1 \big[G_1^e(\vec{q}) e^{i\vec{q}\vec{R}_1} + G_2^g(\vec{q}) e^{i\vec{q}\vec{R}_2}\big] \\ & + n_2 \big[G_1^g(\vec{q}) e^{i\vec{q}\vec{R}_1} + G_2^e(\vec{q}) e^{i\vec{q}\vec{R}_2}\big]\Big\}\end{aligned} \tag{2.61}$$

Here G_i^e, G_i^g are the constants of the electron-phonon coupling in excited and ground states of i-th molecule, $b_{\vec{q}}$, $b^+_{\vec{q}}$ are annihilation and creation operators of phonons with wavevector $\vec{q}$, $\vec{R}_1$ and $\vec{R}_2$ characterize localization of first and second molecules, respectively. Now, we shall perform the transformation eliminating linear in $b_{\vec{q}}$, $b^+_{\vec{q}}$ terms, similar to that of (2.49)

$$\mathcal{H} = \exp(S)\, \hat{\mathcal{H}} \exp(-S) \tag{2.62}$$

where

$$\begin{aligned}S = \sum_{\vec{q}}' \Big\{ & \big[G_1^e(\vec{q}) e^{i\vec{q}\vec{R}_1} + G_2^g(\vec{q}) e^{i\vec{q}\vec{R}_2}\big] n_1 + \\ & + \big[G_1^g(\vec{q}) e^{i\vec{q}\vec{R}_1} + G_2^e(\vec{q}) e^{i\vec{q}\vec{R}_2}\big] n_2 \Big\} (b_{\vec{q}} - b^+_{-\vec{q}})\end{aligned}$$

In the new representation the Hamiltonian obtains the form

$$\mathcal{H} = J_1 n_1 + J_2 n_2 + \sum_{\vec{q}}{}' \hbar\omega_{\vec{q}} \left(b^+_{\vec{q}} b_{\vec{q}} + \frac{1}{2}\right) + r_+ V_{12} \Pi^+ + r_- V_{21} \Pi \tag{2.63}$$

where

$$J_1 = E_1 - \sum_{\vec{q}}{}' \left| G_1^e(\vec{q}) e^{i\vec{q}\vec{R}_1} + G_2^g(\vec{q}) e^{i\vec{q}\vec{R}_2} \right|^2 \hbar\omega_{\vec{q}}$$

$$J_2 = E_2 - \sum_{\vec{q}}{}' \left| G_1^g(\vec{q}) e^{i\vec{q}\vec{R}_1} + G_2^e(\vec{q}) e^{i\vec{q}\vec{R}_2} \right|^2 \hbar\omega_{\vec{q}} \tag{2.64}$$

$$\Pi^+ = \exp\left(\sum_{\vec{q}} \ell(\vec{q}) \left(b_{\vec{q}} - b^+_{-\vec{q}}\right)\right) \tag{2.65}$$

and

$$\ell(\vec{q}) = \ell_1(\vec{q}) e^{i\vec{q}\vec{R}_1} - \ell_2(\vec{q}) e^{i\vec{q}\vec{R}_2} \tag{2.66}$$

$$\ell_1(\vec{q}) = G_1^e(\vec{q}) - G_1(\vec{q})^g, \quad \ell_2(\vec{q}) = G_2^e(\vec{q}) - G_1^g(\vec{q}) \tag{2.67}$$

We see that this Hamiltonian is isomorphous to that of (2.50)-(2.53) describing the electron transfer process in the condensed medium. In the next chapters we will present other examples of problems described by the Hamiltonian isomorphous to that of (2.50).

III. THE EQUATIONS OF MOTION

3.1 Derivation of master equations from von Neumann equation

The electron transfer and energy transfer (Chapters 2.4, 2.5) and corresponding Hamiltonians (2.50) and (2.63) provide us with examples of the typical situation in the theory of rate processes in condensed media. We are dealing with a subsystem interacting with the medium having a continuum of degrees of freedom. It will be convenient to use the following terminology.

We shall call the subsystem we are interested in, a dynamic system, and its surroundings a dissipative system (it is a source of dissipation). The dynamic system has finite number of degrees of freedom and discrete energy levels, while the dissipative system has a continuum of degrees of freedom and continuous energy spectrum. The dynamic and dissipative subsystems, interacting with each other, together form a closed system. The time behavior of this closed system can be described by the von Neumann equation (1.23) for the density matrix ρ of the whole system

$$i\hbar \frac{\partial \rho}{\partial t} = [\mathcal{H}, \rho] \tag{3.1}$$

The Hamiltonian $\mathcal{H}$ may be presented in the form

$$\mathcal{H} = E + F + V + G \tag{3.2}$$

where E - is the Hamiltonian of the dynamic system with the eigenvalues E_n and eigenfunctions $|n\rangle$; F - the Hamiltonian of the dissipative system with eigenvalues F_α and eigenfunctions $|\alpha\rangle$: V - is the interaction energy between dynamic and dissipative systems with matrix elements $V_{m\alpha n\beta}$, and G is the perturbation energy in the dissipative system with matrix elements $G_{\alpha\beta}$. This perturbation energy G is the source of the relaxation processes in the dissipative system itself.

Our final goal is to describe the time behavior of the dynamic subsystem. The density matrix of the whole system in the representation of the unperturbed Hamiltonian

$$\mathcal{H}_0 = E + F \tag{3.3}$$

has matrix elements

$$\rho_{m\alpha, n\beta}$$

where Latin indices relate to the dynamic subsystem and Greek indices to the dissipative subsystem. The behavior of the dynamic subsystem may be described by the density matrix

$$\sigma_{mn} = \sum_{\alpha} \rho_{m\alpha, n\alpha} \quad , \quad \sigma = \mathrm{Tr}\, \rho \tag{3.4}$$

The question arises whether it is possible to find out the equations determining the temporal behavior of the dynamic subsystem only. In other words, the question is whether the von Neumann equation (3.1) (which is valid for the whole subsystem) may be reduced to the equations for density matrix σ. The general answer to this question is negative. The density matrix of the dynamic subsystem at the moment t, $\sigma(t)$, is determined not only by the initial conditions $\sigma(0)$ but also by the initial conditions $\rho(0)$ of the density matrix of the whole system. However, in some special cases, under special physical conditions it is possible to perform approximate reduction to the equations for the subsystem density matrix σ only. Later on we will show how this approximate reduction may be performed.

First, we will consider the relaxation process in the dissipative system, provided there is no interaction with the dynamic subsystem (V=0). In this case the total Hamiltonian of the dissipative system equals

$$\mathcal{H} = F + G \tag{3.4}$$

For example, the role of F may play the Hamiltonian of phonons (1.98), and the role of G may play the unharmonic terms in the Hamiltonian (1.97).

We would try to derive from the basic quantum-mechanical equation (3.1) the master equation for diagonal elements of the density matrix ρ:

$$\dot{\rho}_{\alpha\alpha} = -\sum_{\alpha'} \left(w_{\alpha\alpha'}\, \rho_{\alpha\alpha} - w_{\alpha'\alpha}\, \rho_{\alpha'\alpha'} \right) \tag{3.5}$$

(This equation coincides with (1.99) for the case of relaxation of phonons due to unharmonic interaction terms). As was mentioned, this type of equation was established first by Pauli [3]. In his derivation, Pauli used so-called repeated Random Phase Assumption (RPA). It means that such an equation has been derived for the small interval of time Δt, assuming that at the beginning of the interval the off-diagonal elements of the density matrix

$$\rho_{\alpha\alpha'}(0) = 0 \quad , \quad \alpha \neq \alpha' \tag{3.6}$$

Then, as is clear from the derivation of the Golden Rule Formula (1.80), that for a small enough time interval satisfying (1.78), one can derive the master equation. If at the end of this period Δt one again makes the same random phase assumption (RPA), it is possible to establish the equation (3.5) for an arbitrary moment of time. Of course, such repeated RPA has no direct justification. Van Hove [34] derived the master equations without repeated RPA, using instead of it RPA only at the initial moment t=0.

The name RPA is connected with the fact that off-diagonal elements of the density matrix may describe coherent effects in the quantum system. In the classical systems this coherency is determined by phase relations. We will not repeat here the rather complicated derivation of Van Hove and will use instead the Zwanzig formalism [35].

Let us introduce the operator P which transforms matrix ρ into its diagonal part, i.e.

$$\rho = \rho_1 + \rho_2 , \quad \rho_1 = P\rho , \quad \rho_2 = (1-P)\rho$$
$$\rho_{1\alpha\beta} = \delta_{\alpha\beta}\,\rho_{\alpha\alpha} , \quad \rho_{2\alpha\beta} = (1-\delta_{\alpha\beta})\,\rho_{\alpha\beta} \tag{3.7}$$

We can write the equation (3.1) for the density matrix in the following form

$$\frac{\partial \rho}{\partial t} = -iL\rho = -i\,(L_0 + L_1) \tag{3.8}$$

where

$$L\rho = \hbar^{-1}[\mathcal{H}, \rho] = \hbar^{-1}[F+G, \rho] \equiv L_0\rho + L_1\rho \tag{3.9}$$

The operator L is called the Liouville operator. In general, we will call the operators acting on the density matrix (and not on the wavefunction) the Liouville type operators. According to the definition

$$(L\rho)_{\alpha\beta} = \sum_{\alpha'\beta'} L^{\alpha'\beta'}_{\alpha\beta}\,\rho_{\alpha'\beta'} \tag{3.10}$$

According to (3.7), (3.9) and (3.10)

$$L_{\alpha\beta}^{\alpha'\beta'} = \hbar^{-1}\left(\mathcal{H}_{\alpha\alpha'}\,\delta_{\beta\beta'} - \mathcal{H}_{\beta'\beta}\,\delta_{\alpha'\alpha}\right) \tag{3.11}$$

$$P_{\alpha\beta}^{\alpha'\beta'} = \delta_{\alpha\beta}\,\delta_{\alpha\alpha'}\,\delta_{\beta\beta'} \tag{3.12}$$

For the arbitrary Liouville operators the multiplication rule is held

$$(L_1 L_2)_{\alpha\beta}^{\alpha'\beta'} = \sum_{\alpha''\beta''} L_{\alpha\beta}^{\alpha''\beta''}\, L_{\alpha''\beta''}^{\alpha'\beta'} \tag{3.13}$$

The equation (3.8) can be rewritten in the form

$$\frac{\partial\rho_1}{\partial t} = -iPL(\rho_1+\rho_2)\,, \quad \frac{\partial\rho_2}{\partial t} = -i(1-P)L(\rho_1+\rho_2) \tag{3.14}$$

The latter equation may be also put in the form

$$\frac{\partial\rho_2}{\partial t} + i(1-P)L\rho_2 = -i(1-P)L\rho_1 \tag{3.15}$$

The solution of this equation may be presented as the sum of the general solution of the equation

$$\frac{\partial\rho_2}{\partial t} + i(1-P)L\rho_2 = 0 \tag{3.16}$$

and the specific solution of the equation (3.15). The general solution of (3.16) has the form

$$\tilde{\rho}_2 = \exp\left[-it(1-P)L\right]\rho_2(0) \tag{3.17}$$

and the specific solution of (3.15) may be taken as

$$\tilde{\tilde{\rho}}_2 = -i\int_0^t d\tau \exp\left[-i\tau(1-P)L\right](1-P)L\rho_1(t-\tau)$$

Thus the equation for diagonal elements of the density matrix ρ_1 obtains the form

$$\frac{\partial\rho_1}{\partial t} = -iPL\exp\left[-it(1-P)L\right]\rho_2(0) - iPL\rho_1 - \\ - \int_0^t d\tau\, K(\tau)\rho_1(t-\tau) \tag{3.18}$$

where

$$K(\tau) = PL\exp\left[-i\tau(1-P)L\right](1-P)L \tag{3.19}$$

Therefore, in the general case, no closed equation is obtained for ρ_1, since, as can be seen from (3.18), the behavior of ρ_1 at the time t is determined not only by ρ_1 at the time t = 0 but also by $\rho_2(0)$.

Now we will show that if we assume $\rho_2(0) = 0$, i.e. we make RPA at the initial moment of time, then the equation (3.18) may be reduced to the master equation (3.5) provided we make certain approximations. First, let us make some identical transformations. Taking into account the assumption (3.6) $\rho_2(0) = 0$, we would rewrite the equation (3.18) in the form

$$\frac{\partial\rho_{\alpha\alpha}}{\partial t} = -i\sum_\beta (PL)_{\alpha\alpha}^{\beta\beta}\rho_{\beta\beta} - \sum_\beta{}' \int_0^t K_{\alpha\alpha}^{\beta\beta}(\tau)\rho_{\beta\beta}(t-\tau)\,d\tau$$

From (3.12), (3.13) and (3.19) it follows that

$$(PL)^{\beta\beta}_{\alpha\alpha} = L^{\beta\beta}_{\alpha\alpha} = \hbar^{-1}\left(\mathcal{H}_{\alpha\beta}\,\delta_{\alpha\beta} - \mathcal{H}_{\beta\alpha}\,\delta_{\beta\alpha}\right) = 0\,,$$

$$K^{\beta\beta}_{\alpha\alpha} = \sum_{\substack{\alpha'\beta'\\ \alpha''\beta''}} P^{\alpha\alpha}_{\alpha\alpha}\, L^{\alpha'\beta'}_{\alpha\alpha}\left\{\exp\left[-i\tau(1-P)L\right]\right\}^{\alpha''\beta''}_{\alpha'\beta'}\left(1-\delta_{\alpha''\beta''}\right) L^{\beta\beta}_{\alpha''\beta''}$$

(3.20)

$$L^{\alpha'\beta'}_{\alpha\alpha} = \hbar^{-1}\left(\mathcal{H}_{\alpha\alpha'}\,\delta_{\alpha\beta'} - \mathcal{H}_{\beta'\alpha}\,\delta_{\alpha\alpha'}\right)$$

$$L^{\beta\beta}_{\alpha''\beta''} = \hbar^{-1}\left(\mathcal{H}_{\alpha''\beta}\,\delta_{\beta''\beta} - \mathcal{H}_{\beta\beta''}\,\delta_{\beta\alpha''}\right)$$

(3.21)

As is clear from these expressions for L, they contain only the off-diagonal elements of H, i.e. the matrix elements of the perturbation energy G. It means that if we assume the smallness of this perturbation G and would calculate the matrix K up to the terms G^2, then we may put in the exponent instead of L its approximate value L_o, and

$$K = PL_1 \exp\left[-i\tau(1-P)L_0\right](1-P)L_1 \qquad (3.22)$$

In this approximation the explicit expression for the matrix element of K is equal according to (3.20)-(3.22)

$$K^{\beta\beta}_{\alpha\alpha}(\tau) = \hbar^{-2} \sum_{\alpha' \neq \alpha}' |G_{\alpha\alpha'}|^2 \left(e^{i\omega_{\alpha\alpha'}\tau} + e^{-i\omega_{\alpha\alpha'}\tau}\right)\delta_{\alpha\beta}$$

$$-\hbar^{-2}|G_{\alpha\beta}|^2\left(e^{i\omega_{\alpha\beta}\tau} + e^{-i\omega_{\alpha\beta}\tau}\right)\left(1-\delta_{\alpha\beta}\right) \tag{3.23}$$

Thus the equation for the diagonal elements of the density matrix obtains the form

$$\frac{\partial \rho_{\alpha\alpha}}{\partial t} = -\sum_{\beta}' \int_0^t d\tau\, R_{\alpha\beta}(\tau)\left[\rho_{\alpha\alpha}(t-\tau) - \rho_{\beta\beta}(t-\tau)\right] \tag{3.24}$$

where

$$R_{\alpha\beta}(t) = \hbar^{-2}|G_{\alpha\beta}|^2\left(e^{i\omega_{\alpha\beta}t} + e^{-i\omega_{\alpha\beta}t}\right) \tag{3.25}$$

The equations (3.24), (3.25) are not master equations (3.5) yet. In the master equations (3.5) the derivative of the probability $\rho_{\alpha\alpha}(t)$ of finding a system in the state α at time t is determined by the probabilities $\rho_{\alpha'\alpha'}(t)$ at the same moment and do not depend on the prehistory of the system. Such an approximation is sometimes called the Markovian approximation. The equations (3.24) contain the memory about the state of the system in all previous moments of time. Such behavior of the system is called non-Markovian.

We will show, that essentially in the same approximation at which the notion of transition probability per unit time may be introduced (see Chapter 1.3), the equations (3.24) may be reduced to the master equations (3.5). For this purpose the r.h.s. of (3.24) will be rewritten with the aid of the identical transformation as

$$\frac{\partial \rho_{\alpha\alpha}}{\partial t} = -\sum_{\beta}' \frac{1}{2\pi i} \lim_{\varepsilon \to +0} \int_{\varepsilon - i\infty}^{\varepsilon + i\infty} \hat{R}_{\alpha\beta}(p)\left[\hat{\rho}_{\alpha\alpha}(p) - \hat{\rho}_{\beta\beta}(p)\right] e^{pt}\, dp \tag{3.26}$$

where $\hat{R}(p)$, $\hat{\rho}_{\beta\beta}(p)$ are the Laplace transforms of the functions $R_{\alpha\beta}(t)$, $\rho_{\beta\beta}(t)$.

In the explicit form the equation (3.26) obtains the form

$$\frac{\partial \rho_{\alpha\alpha}}{\partial t} = -\frac{1}{2\pi i} \lim_{\varepsilon\to+0} \int_{\varepsilon-i\infty}^{\varepsilon+i\infty} \sum_{\beta}{}' |G_{\alpha\beta}|^2 \left(\frac{1}{p-i\omega_{\alpha\beta}} + \frac{1}{p+i\omega_{\alpha\beta}} \right) \left[\hat{\rho}_{\alpha\alpha}(p) - \hat{\rho}_{\beta\beta}(p) \right] e^{pt} dp \tag{3.27}$$

Now, we will make an assumption that the energy spectrum of the dissipative subsystem F_α has a (quasi) continuous character and the summation over β includes integration over frequencies $\omega_{\alpha\beta}$. Then we can introduce two characteristic frequency intervals. One of them

$$|\Delta p| \sim T_{diss}^{-1} \tag{3.28}$$

characterizes relaxation of the probabilities $\rho_{\beta\beta}(t)$. If T_{diss} is the relaxation time of the quantities $\rho_{\beta\beta}(t)$, then (3.28) characterizes the interval of p near $p \approx 0$, where $\hat{\rho}_{\beta\beta}(p)$ is essentially non-zero. Another frequency interval characterizes the sum (or integral) over $\omega_{\alpha\beta}$ in (3.27). We will designate the characteristic frequency

$$\omega^* = \tau_c^{-1} \tag{3.29}$$

According to the definition, when

$$|p| \ll \omega^*$$

one may neglect p-dependence of $\hat{R}(p)$ in the r.h.s. (3.26), (3.27). Now, it is clear that if

$$\omega^* \gg T_{diss}^{-1} \tag{3.30}$$

the interval of the essential variation of $\hat{\rho}(p)$ ($\approx T_{diss}^{-1}$) is much smaller than that of the sum $\hat{R}_{\alpha\beta}(p)$ over $\beta(\approx\omega^*)$. In this case $\hat{R}_{\alpha\beta}(p)$ in the r.h.s. (3.26), (3.27) may be put outside the integral with the meaning of p close to zero:

$$\frac{\partial\rho_{\alpha\alpha}}{\partial t} = -\sum_{\beta}{}' \lim_{\varepsilon\to+0} \hat{R}_{\alpha\beta}(\varepsilon)\int_{\varepsilon-i\infty}^{\varepsilon+i\infty}\left[\hat{\rho}_{\alpha\alpha}(p)-\hat{\rho}_{\beta\beta}(p)\right]e^{pt}\,dp =$$

$$= -\sum_{\beta}{}' \lim_{\varepsilon\to+0} \hat{R}_{\alpha\beta}(\varepsilon)\left[\rho_{\alpha\alpha}(t)-\rho_{\beta\beta}(t)\right] \tag{3.31}$$

This equation already has the form of the master equation (3.5) and may be rewritten in the form

$$\frac{\partial\rho_{\alpha\alpha}}{\partial t} = -\sum_{\beta} w_{\alpha\beta}\left(\rho_{\alpha\alpha}-\rho_{\beta\beta}\right) \tag{3.32}$$

where

$$w_{\alpha\beta} = \lim_{\varepsilon\to+0} \hat{R}_{\alpha\beta}(\varepsilon) = \frac{2\pi}{\hbar}\left|G_{\alpha\beta}\right|^2 \delta\left(F_\alpha - F_\beta\right) \tag{3.33}$$

It is clear that the characteristic relaxation time T_{diss} is determined by the relation

$$T_{diss}^{-1} \simeq \frac{2\pi}{\hbar}\sum_{\beta}\left|G_{\alpha\beta}\right|^2 \delta\left(F_\alpha - F_\beta\right) \tag{3.34}$$

and that our basic assumption (3.30) essentially coincides with the relation (1.79) determining the possibility of an introduction of the transition probability per unit time. The relation (3.30) together with (3.34) exhibits in an explicit way, the meaning of the smallness of the perturbation energy G. The assumption about the smallness of G was made in transition from (3.18) to (3.22), when L in the exponent (3.18) was replaced by L_o.

It is worthwhile to mention that the steady-state solution of the master equation (3.32) corresponds to the micro-canonical ensemble

$$\rho_{\alpha\alpha} = \rho_{\beta\beta} \quad \text{for} \quad F_\alpha = F_\beta \tag{3.35}$$

i.e. the equipartition over all states with the same energy takes place.

Comment on the RPA at initial time

The irreversible behavior of the system and description of it by the master equations have been derived from the first principles, i.e. from the von Neumann equation (3.1), but for the specific initial condition (3.6)

$$\rho_2(0) = 0 \tag{3.36}$$

corresponding to RPA at the initial time $t = 0$. Of course, such an assumption is not a very satisfactory one. The moment of time $t = 0$ is not singled out, and at another moment of time $t_o \neq 0$, $\rho_2(t_o)$ is already not zero (see (3.17)). So, we will try to proceed without making assumption (3.36). Assuming again the smallness of the perturbation energy G and expanding the r.h.s. of the exact equation (3.17) up to the terms of order G^2 we get

$$\frac{\partial \rho_{\alpha\alpha}}{\partial t} = -i\hbar^{-1} \sum_{\gamma \neq \alpha} \left[G_{\alpha\gamma} \rho_{\gamma\alpha}(0) e^{-i\omega_{\gamma\alpha} t} - \rho_{\alpha\gamma}(0) G_{\gamma\alpha} e^{i\omega_{\gamma\alpha} t} \right]$$
$$+ O(G^2)\rho_2(0) - \sum_{\beta} \int_0^t R_{\alpha\beta}(\tau) \left[\rho_{\alpha\alpha}(t-\tau) - \rho_{\beta\beta}(t-\tau) \right] d\tau \tag{3.37}$$

The term of second order of G - $O(G^2)$ in the r.h.s. has the same structure as the first sum in the r.h.s. of (3.37).

The energy spectrum of the dissipative system is supposed to be continuous (or quasi-continuous). It means that the sums in the r.h.s. of (3.37) may be replaced by the Fourier integrals over the frequencies $\omega_{\alpha\beta}$. These sums are of types:

$$\sum_{\gamma} G_{\alpha\gamma} \rho_{\gamma\alpha}(0) e^{-i\omega_{\gamma\alpha} t} \,, \quad \sum_{\beta} |G_{\alpha\beta}|^2 e^{i\omega_{\alpha\beta}\tau} \rho_{\beta\beta}(t-\tau)$$

Each of these terms may be characterized by corresponding characteristic frequency scales. Let us designate by ω^* the minimal of all these scales, and let us assume that the characteristic relaxation time T_{diss} of the dissipative system (3.34) is much larger than $\tau_c = \omega^{*-1}$

$$\omega^* \gg T_{diss}^{-1} \tag{3.38}$$

This condition is similar to that of (3.30). Then we may neglect the first two terms in (3.37). They would vanish in time $t \gtrsim \tau_c = \omega^{*-1}$ which is much smaller than the relaxation time T_{diss}. In this approximation we get master equation (3.32) without Random Phase ASsumption (RPA) at the initial time.

It is worth-while to mention that in his paper [34] van Hove has stressed that master equations may be derived for the specific class of initial condition, namely, for the wavepockets with

$$\Delta\omega \gg T_{diss}^{-1}$$

When all the above mentioned frequency scales (for each term in the r.h.s. of (3.37)) are of the same order, this condition is not an additional limitation. It is the necessary condition or derivation of master equations. In the opposite case

$$\Delta\omega \lesssim T_{diss}^{-1}$$

not only the first terms would not vanish quickly enough, but the last term on the r.h.s. of (3.37) cannot be reduced to the last term of (3.32), i.e., to the master equation. In this case the variation of $R_{\alpha\beta}(t)$ in the sum (3.24) and $\rho_{\beta\beta}(t-\tau)$ are of the same order; the Markovian approximation is not valid and memory effects are contained in all terms of the r.h.s. of (3.37).

3.2 Equations for density matrix of dynamic subsystem

As has been already mentioned in the theory of rate processes in condensed media, we are dealing with a subsystem interacting with its surroundings - the condensed medium. We will be interested in the time behavior of the subsystem. Generally speaking, it is determined by the von Neumann equation (3.1) for the whole system (subsystem+condensed medium) with the Hamiltonian (3.2). Under certain physical conditions it is possible to perform an approximate reduction of the equation (3.1) to the equations for the subsystem density matrix σ_{mn} (3.4) [36]. In this section we will show how this approximate reduction may be performed.

For this purpose we will introduce new projection operator D (instead of (3.7)) dividing the density matrix of the whole system into two parts

$$\rho = \rho_1 + \rho_2 \equiv D\rho + (1-D)\rho \tag{3.39}$$

where

$$\rho_1 = \rho_{m\alpha n\alpha}\,, \quad \rho_2 = \rho_{m\alpha n\beta}(1-\delta_{\alpha\beta}) \tag{3.40}$$

and thus operator D is defined as

$$\mathcal{D}_{m\alpha n\beta}^{m'\alpha' n'\beta'} = \delta_{\alpha\beta}\,\delta_{mm'}\,\delta_{\alpha\alpha'}\,\delta_{nn'}\,\delta_{\beta\beta'} \tag{3.41}$$

According to (3.1) and (3.2) the von Neumann equation may be rewritten as

$$\frac{\partial\rho}{\partial t} = -iL\rho \tag{3.42}$$

$$L = L_0 + L^V + L^G \equiv L_0 + L_1$$

$$L_0\rho = \hbar^{-1}[E+F,\rho]\ ,\quad L^V\rho = \hbar^{-1}[V,\rho] \tag{3.43}$$

$$L^G\rho = \hbar^{-1}[G,\rho]$$

Assuming that condition (3.38) (valid for the dissipative system) is fulfilled, we may rewrite the equation (3.42) as

$$\frac{\partial\rho_1}{\partial t} = -i\mathcal{D}L_0\rho_1 - i\mathcal{D}L^V\rho_1 - \int_0^t K^{GG}(\tau)\,\rho_1(t-\tau)\,d\tau$$

$$-\int_0^t K^{VV}(\tau)\,\rho_1(t-\tau)\,d\tau - \int_0^t K^{GV}(\tau)\,\rho_1(t-\tau)\,d\tau$$

(3.44)

where

$$K^{VV} = DL^{V}\exp[-i\tau(1-D)L_{o}](1-D)L^{V}$$

$$K^{GG} = DL^{G}\exp[-i\tau(1-D)L_{o}](1-D)L^{G} \qquad (3.45)$$

$$K^{GV} = DL^{V}\exp[-i\tau(1-D)L_{o}](1-D)L^{G} +$$
$$+ DL^{G}\exp[-i\tau(1-D)L_{o}](1-D)L^{V}$$

Here it has been assumed that $V + G$ is small perturbation and thus L in the exponent (3.45) may be replaced by L_o.

The equation (3.44) determines time evolution of the density matrix and in general cannot be reduced to the equation for the density matrix (3.4)

$$\sigma_{mn} = \sum_{\alpha} \rho_{m\alpha n\alpha}$$

However, under specific conditions such a reduction may be approximately performed. These conditions corresponds to more rapid relaxation in the dissipative system itself in comparison iwth the relaxation of the dynamic subsystem. Namely, we assume that

$$T_{diss}^{-1} \gg T_{dyn}^{-1} \qquad (3.46)$$

where T_{dyn} is the relaxation time of the dynamic subsystem. We also assume that the relaxation process in the dissipative system is governed by the master equation (3.32).

In the explicit form the equation (3.44) may be written as

$$\frac{\partial \rho_{m\alpha n\alpha}}{\partial t} = -i\hbar^{-1} \sum_{n} \left(V_{m\alpha k\alpha} \rho_{k\alpha n\alpha} - \rho_{m\alpha k\alpha} V_{k\alpha n\alpha} \right) -$$

$$-i\hbar^{-1} (E_m - E_n) \rho_{m\alpha n\alpha} - \sum_{\beta}{}' w_{\alpha\beta} \left(\rho_{m\alpha n\alpha} - \rho_{m\beta n\beta} \right) - \tag{3.47}$$

$$-\sum_{m'n'\beta}{}' \int_0^t \left[\left(K(\tau) \right)^{VV\; m'\beta n'\beta}_{m\alpha n\alpha} + \left(K(\tau) \right)^{GV\; m'\beta n'\beta}_{m\alpha n\alpha} \right] \rho_{m'\beta n'\beta}(t-\tau)\, d\tau$$

The condition (3.46) is essentially equivalent to the condition

$$\lambda = \frac{|V|}{|G|} \ll 1 \tag{3.48}$$

This condition allows us to seek the solution of the equation (3.47) in the form

$$\rho_{m\alpha n\alpha} = \sigma_{mn} P_{\alpha} + \eta_{mn\alpha}, \quad \sum_{\alpha} \eta_{mn\alpha} = 0 \tag{3.49}$$

where $P_\alpha = \rho_{\alpha\alpha}$ is assumed to be the equilibrium solution of the equation (3.32). The last equation (3.49) follows from the normalization conditions

$$\sum_{\alpha}{}' \rho_{m\alpha n\alpha} = \sigma_{mn}, \quad \sum_{\alpha}{}' P_{\alpha} = 1$$

Substituting (3.49) for (3.47) we get the equation for $\eta_{mn\alpha}$

$$\dot{\eta}_{mn\alpha} + i\omega_{mn}\eta_{mn\alpha} + i\hbar^{-1}\left(V_{m\alpha k\alpha}\eta_{k\alpha n\alpha} - \eta_{m\alpha k\alpha}V_{k\alpha n\alpha}\right) +$$

$$+ \sum_{\beta} w_{\alpha\beta}\left(\eta_{mn\alpha} - \eta_{mn\beta}\right) +$$

$$+ \sum_{m'n'\alpha'} \int_0^t \left[\left(K(\tau)\right)^{VV\, m'\alpha' n'\alpha'}_{m\alpha n\alpha} + \left(K(\tau)\right)^{VG\, m'\alpha' n'\alpha'}_{m\alpha n\alpha}\right] \eta_{m'n'\alpha'}(t-\tau) = \qquad (3.50)$$

$$= -\dot{\sigma}_{mn} P_\alpha - i\omega_{mn}\sigma_{mn} P_\alpha - i\hbar^{-1} P_\alpha \sum_{n} \left(V_{m\alpha k\alpha}\sigma_{kn} - \sigma_{mk} V_{k\alpha n\alpha}\right) -$$

$$- \sum_{m'n'} \int_0^t \left[\left(K(\tau)\right)^{VV\, m'\alpha' n'\alpha'}_{m\alpha n\alpha} + \left(K(\tau)\right)^{GV\, m'\alpha' n'\alpha'}_{m\alpha n\alpha}\right] P_\alpha \sigma_{m'n'}(t-\tau)\, d\tau$$

Here

$$\omega_{mn} = \left(E_m - E_n\right)/\hbar$$

Performing the transformation

$$\sigma_{mn}, \eta_{mn\alpha} \rightarrow \bar{\sigma}_{mn} e^{-i\omega_{mn}t}, \bar{\eta}_{mn\alpha} e^{-i\omega_{mn}t}$$

it is easy to verify that the terms η have the order of magnitude

$$\eta \sim \left(T_{diss}/T_{dyn}\right)\sigma P, \quad \eta/\sigma P \ll 1 \qquad (3.51)$$

This condition being satisfied, after summation over α in (3.50) and neglecting terms with η, one can easily get the equation for the density matrix of the dynamic subsystem

$$\dot{\sigma} = -i\hbar^{-1}\left[E+\bar{V},\sigma\right] - \int_0^t \left[\bar{K}^{VV}(\tau) + \bar{K}^{VG}(\tau)\right]\sigma(t-\tau)\,d\tau \tag{3.52}$$

where

$$\bar{V}_{mn} = \sum_{\alpha} P_{\alpha} V_{m\alpha n\alpha} \tag{3.53}$$

$$\bar{K}_{mn}^{m'n'}(\tau) = \sum_{\alpha\alpha'} K(\tau)_{m\alpha n\alpha}^{m'\alpha' n'\alpha'} P_{\alpha} \tag{3.54}$$

In the explicit form the matrix elements of $\bar{K}^{VV}$, $\bar{K}^{GV}$ may be rewritten as

$$\begin{aligned}\hbar^2\left(\bar{K}^{VV}(\tau)\right)_{mn}^{m'n'} = {} & \delta_{nn'}\sum_{m_1}\langle \tilde{V}_{mm_1}(\tau)\tilde{V}_{m_1m'}(0)\rangle e^{-i\omega_{m_1n}\tau} + \\ & + \delta_{mm'}\sum_{m_1}\left\langle \tilde{V}_{n'm_1}(0)\tilde{V}_{m_1n}(\tau)\right\rangle e^{-i\omega_{mm_1}\tau} \\ & - \langle \tilde{V}_{n'n}(0)\tilde{V}_{mm'}(\tau)\rangle e^{-i\omega_{m'n}\tau} - \langle V_{n'n}(\tau)V_{mm'}(0)\rangle e^{-i\omega_{mn'}\tau}\end{aligned} \tag{3.55}$$

and

$$\begin{aligned}\hbar^2\left(\bar{K}^{GV}\right)_{mn}^{m'n'} = {} & \delta_{nn'}\left\langle\left[\tilde{V}_{mm'}(\tau),\tilde{W}(0)\right]\right\rangle e^{-i\omega_{m'n}\tau} + \\ & + \delta_{mm'}\left\langle\left[\tilde{W}(0),\tilde{V}_{n'n}(\tau)\right]\right\rangle e^{-i\omega_{mn'}\tau}\end{aligned} \tag{3.56}$$

Here $\langle\ldots\rangle$ means averaging over the thermal equilibrium state of the dissipative system, dependence of the matrix elements on time is determined by the unperturbed Hamiltonian F of the dissipative system, and tilde $\sim$ means that only off-diagonal elements of the matrixes V and W are taken into account

$$\tilde{V}_{m\alpha n\beta} = V_{m\alpha n\beta}\left(1-\delta_{\alpha\beta}\right), \quad \tilde{W}_{\alpha\beta} = W_{\alpha\beta}\left(1-\delta_{\alpha\beta}\right) \tag{3.57}$$

The equation (3.52) may be essentially simplified in the Markovian approximation, i.e. when the variation of $\sigma(t)$ in (3.52) is much slower than that of $\bar{K}^{VV}(\tau)$ and $\bar{K}^{VG}(\tau)$. Quantitatively this approximation means that the correlation time τ_{corr} characterizing the correlation functions (3.55), (3.56) should satisfy the condition

$$\tau_{corr} \ll T_{dyn}, \ \hbar/|V_{nm}| \tag{3.58}$$

where T_{dyn} is the relaxation time of the dynamic subsystem.

Then, using considerations similar to those of preceding sections, one can get

$$\dot{\sigma} = -i\hbar^{-1}\left[E+\bar{V}^{ef}, \sigma\right] - R\sigma \tag{3.59}$$

where E is the Hamiltonian of the dynamic subsystem (see (3.2)),

$$\bar{V}^{ef} = \bar{V} + i\hbar^{-1}\int_0^\infty \langle [V(\tau), G(0)] \rangle \, d\tau \tag{3.60}$$

and supermatrix R is defined by the expression

$$R_{mn}^{m'n'} = \int_0^\infty \left(\bar{K}^{VV}(\tau)\right)_{mn}^{m'n'} e^{i\omega_{m'n'}\tau} \, d\tau \tag{3.61}$$

3.3 Master equations for dynamic subsystem

Now we will derive conditions under which the equations for the density matrix of the dynamic subsystem (3.59) may be reduced to the master equations for the dynamic subsystem. For this purpose we will transform (3.59) to the interaction representation

$$\sigma_{mn} = \bar{\sigma}_{mn} e^{-i\omega_{mn}t} \tag{3.62}$$

and get the equation (3.59) in the form

$$\dot{\sigma}_{mn} = -i\hbar^{-1}\sum_k{}' \left(\bar{V}^{ef}_{mk} e^{i\omega_{mk}t}\bar{\sigma}_{kn} - \bar{\sigma}_{mk}\bar{V}^{ef}_{kn} e^{i\omega_{kn}t}\right) - \tag{3.63}$$

$$-\sum_{m'n'} R^{m'n'}_{mn}\bar{\sigma}_{m'n'} e^{i(\omega_{mn}-\omega_{m'n'})t}$$

Now we will make the following assumption:

(a) All the levels of the dynamic subsystem are non-degenerate:

$$\omega_{mn} \neq 0 \, , \quad \text{if} \quad m \neq n$$

(b)

$$|\omega_{mn}| \gg \frac{|V^{ef}_{mn}|}{\hbar} \, , \quad T^{-1}_{dyn} \tag{3.64}$$

where T_{dyn}-characteristic time of relaxation having the order of magnitude

$$\left| R^{m'n'}_{mn} \right|^{-1}$$

Now, the equations (3.63) contain terms which do not depend on time explicitly (e.g. term with $\omega_{mn} = \omega_{m'n'}$ in the r.h.s. of (3.63)) and rapidly varying terms ($\propto e^{i\omega_{mn}t}$). It is easy to show that contribution of these rapidly varying terms is much less than those which are independent of time.*)

In this approximation the equations for the diagonal elements σ_{nn} of the density matrix obtain the form

$$\dot{\sigma}_{nn} = -\sum_k{}' R^{kk}_{nn} \sigma_{kk} \tag{3.65}$$

Using (3.61) and (3.65) we get

$$R^{nn}_{nn} = \hbar^{-2} \sum_{k \neq n}{}' \int_{-\infty}^{\infty} \langle \tilde{V}_{nk}(\tau) \tilde{V}_{kn}(0) \rangle e^{-i\omega_{kn}\tau} d\tau =$$

$$= \sum_{k \neq n}{}' \sum_{\substack{\alpha,\alpha' \\ \alpha \neq \alpha'}}{}' P_\alpha \frac{2\pi}{\hbar} |V_{n\alpha k\alpha'}|^2 \delta(F_\alpha - F_{\alpha'} - E_k + E_n) = \sum_{k \neq n}{}' w_{nk}$$

$$\underset{(k \neq n)}{R^{kk}_{nn}} = -\hbar^{-2} \int_{-\infty}^{\infty} \langle \tilde{V}_{kn}(\tau) \tilde{V}_{nk}(0) \rangle e^{-i\omega_{nk}\tau} d\tau = -w_{kn}$$

---

*We will show this on a very simple example which may easily be extended. Let us examine the equation

$$\dot{\sigma} + R\sigma = R_1 \sigma e^{i\omega t} \tag{1}$$

Though it may be solved exactly we will seek an approximate solution of this equation in the form

$$\sigma = \bar{\sigma} + \tilde{\sigma} \tag{2}$$

where $\bar{\sigma}$ satisfies the equation (1) without r.h.s. Substituting (2) into (1) we get

$$\dot{\tilde{\sigma}} + R\tilde{\sigma} = R_1 \bar{\sigma} \exp(i\omega t) + R_1 \tilde{\sigma} \exp(i\omega t)$$

Assuming that last term in the r.h.s. is small we obtain an estimate for $\tilde{\sigma}$

$$\sigma \sim \frac{R_1}{i\omega + R} \bar{\sigma} e^{i\omega t} \sim \frac{R_1}{\omega} \bar{\sigma}$$

Thus if $R_1 \ll \omega$, the contribution of the oscillating term is also small.

Here $P_\alpha = \rho_{\alpha\alpha}$ are diagonal elements of the equilibrium density matrix of the dissipative subsystem. As it has been shown (3.35), these matrix elements correspond to the micro-canonical ensemble. Having in mind the well known fact that the averaging with the aid of the microcanonical ensemble is equivalent to that of the one with the aid of the canonical ensemble, we can use for P_α

$$P_\alpha = e^{-F_\alpha/k_B T} \Big/ \sum_\beta{}' e^{-F_\beta/k_B T} \tag{3.67}$$

Now, we see that the equations (3.65) coincide with the master equations for the dynamic system

$$\dot{\sigma}_{nn} = -\sum_k{}' \left(w_{nk}\sigma_{nn} - w_{kn}\sigma_{kk} \right) \tag{3.68}$$

with the coefficients w_{nk} having the meaning of the transition probabilities per unit time (1.80) averaged over the thermal equilibrium ensemble (3.67) of the dissipative system

$$\begin{aligned} w_{nk} &= \frac{2\pi}{\hbar} \sum_{\substack{\alpha\alpha' \\ \alpha\neq\alpha'}} P_\alpha \left| V_{n\alpha,k\alpha'} \right|^2 \delta\left(F_\alpha - F_{\alpha'} - E_k + E\right) = \\ &= \hbar^{-2} \int_{-\infty}^{\infty} \langle \tilde{V}_{nk}(\tau)\tilde{V}_{kn}(0)\rangle e^{-i\omega_{kn}\tau} d\tau \end{aligned} \tag{3.69}$$

From this formula and (3.67) we obtain the relation (1.85) between the transition probabilities $n \rightarrow k$ and $k \rightarrow n$

$$w_{nk} = w_{kn} \exp\left[-(E_n - E_k)/k_B T\right] \tag{3.70}$$

This relation ensures the Boltzmann distribution (1.48) to be the asymptotic solution ($t \to \infty$) of the master equations (3.68). Simple equations for the off-diagonal elements σ_{mn} may be derived if one makes an additional assumption about non-degeneracy of the frequencies ω_{mn}:

$$\omega_{mn} \neq \omega_{m'n'} \quad , \quad \text{if} \quad (m,n) \neq (m',n')$$

and that

$$|\omega_{mn} - \omega_{m'n'}| \gg |\bar{V}^{ef}|/\hbar, \; T_{dyn}^{-1}; \quad (m,n) \neq (m',n') \tag{3.71}$$

Then using almost the same arguments as in the derivation of the master equation (55), we get

$$\dot{\sigma}_{mn} + i\hbar^{-1}\left[E + \bar{V}^{ef} + \Gamma, \sigma\right]_{mn} = \begin{cases} -\sum_k \left(w_{nk}\sigma_{nn} - w_{kn}\sigma_{kk} \right) \\ -\frac{1}{\tau_{mn}}\sigma_{mn} \quad (m \neq n) \end{cases} \tag{3.72}$$

where

$$\tau_{mn}^{-1} = \frac{1}{2}\sum_k \left(w_{nk} + w_{kn} \right) + \frac{\pi}{\hbar}\sum_{\substack{\alpha,\alpha' \\ \alpha\neq\alpha'}} P_\alpha \left| V_{m\alpha' m\alpha} - V_{n\alpha' n\alpha} \right|^2 \delta(E_\alpha - E_{\alpha'})$$

$$\Gamma_{mk} = \hbar^{-1}\sum_{\substack{l,\alpha,\alpha' \\ \alpha\neq\alpha'}} P_\alpha \frac{V_{m\alpha l\alpha'} V_{l\alpha' m\alpha}}{\omega_{lm} + \omega_{\alpha\alpha'}} \delta_{mk} \tag{3.73}$$

The equations (3.72) essentially coincide with generalized Bloch equations (see [37,38,5]). It is worthwhile to stress again that these equations (3.72) are valid for the case of nondegenerate energies (frequencies) of the dynamic subsystem. In the case of the degeneracy we should deal with general equations (3.59) (in Markovian limit). Examples of description of systems with degenerate energy levels will be presented below.

IV. CALCULATION OF RATE COEFFICIENTS IN VARIOUS TEMPERATURE REGIONS

4.1 Equation of motion describing electron transfer

Now, we can apply the general theory developed in the preceding chapter to concrete systems, or, more specifically, to concrete models. In Chapter II we have considered a number of model systems. Among these models were those taking into account interaction with condensed medium. The Hamiltonians (2.47) and (2.61) describe models of electron transfer and energy transfer in condensed media. Since these Hamiltonians are almost isomorphous it is enough to consider the Hamiltonian (2.47) describing the electron transfer in condensed medium.

Taking into account the perturbation energy G in the dissipative system itself (see (3.2)) we can write this Hamiltonian in the form

$$\mathcal{H} = n_1 E_1 + n_2 E_2 + \frac{1}{2}\sum_k^{N}{}' (p_k^2 + \omega_k^2 q_k^2) - n_1 \sum_k^{N}{}' \omega_k^2 q_{1k}^0 q_k - n_2 \sum_k^{N}{}' \omega_k^2 q_{2k}^0 q_k + r_+ V_{12} + r_- V_{21} + G(q_1, q_N) \quad (4.1)$$

Now, we we will again perform the transformation (2.49) eliminating the terms of $-n_i \Sigma \omega_k^2 q_{ik}^0 q_k$ type:

$$U = \prod_k \exp\left[-i\hbar^{-1}(q_{1k}^0 n_1 + q_{2k}^0 n_2) p_k\right] \quad (4.2)$$

As we know, (see (2.50)), the result of this transformation may be written in the form

$$U^+ \mathcal{H} U \to \mathcal{H} = \frac{1}{2}\sum_k{}' (p_k^2 + \omega_k^2 q_k^2) + n_1 J_1 + n_2 J_2 + r_+ V \Pi^+ + r_- V \Pi + G'$$

where all the quantities are defined in Chapter II((2.54) and (2.48))

$$\Pi = \prod_k \exp\left[-\gamma_k (a_k^+ - a_k)\right] \tag{4.3}$$

$$\gamma_k = (\omega_k/2\hbar)^{1/2} (q_{1k}^0 - q_{2k}^0) \tag{4.4}$$

and

$$G' = U^+ G U = G(\ldots, q_k - (n_1 q_{1k}^0 + n_2 q_{2k}^0), \ldots) =$$

$$= G(q_1, \ldots, q_k, \ldots) + \Delta G \tag{4.5}$$

Here ΔG is the contribution of the electronic subsystem into the perturbation energy G. Further on we will assume that the characteristic coupling energy of the condensed medium (the dissipative system) ε is much larger than the energy corresponding to the electronic reorganization, the so-called reorganization energy

$$E_r = \frac{1}{2} \sum_k \omega_k^2 (q_{1k}^0 - q_{2k}^0)^2 \tag{4.6}$$

It can be shown that one can neglect $\Delta G = G' - G$, provided

$$E_r \ll \mathcal{E} \tag{4.7}$$

In this case the transformed Hamiltonian describing the electron transfer may be presented in the form (3.2)

$$\mathcal{H} = n_1 J_1 + n_2 J_2 + \frac{1}{2} \sum_k (p_k^2 + \omega_k^2 q_k^2) + \tag{4.8}$$

$$+ r_+ V \Pi^+ + r_- V \Pi + G$$

where $V_{12} = V_{21} = V$.

Now we can apply the theory developed above (Chapter III) to the model Hamiltonian (4.8). In this case the dynamic system is the two-state system (two electronic states) represented by simple Hamiltonian

$$E = n_1 J_1 + n_2 J_2 = \frac{1}{2}(J_1 + J_2) + r_3 (J_1 - J_2) \tag{4.9}$$

and the dissipative system is a "phonon bath" - set of harmonic oscillator Hamiltonians describing all vibrational modes of the system

$$F = \frac{1}{2}\sum_k{}' (p_k^2 + \omega_k^2 q_k^2) = \sum_k{}' (a_k^+ a_k + \frac{1}{2})\hbar\omega_k \tag{4.10}$$

The equations for 2x2 density matrix σ_{ik} of the dynamic system obtain a comparatively simple form

$$\dot{\sigma}_{11} = -\dot{\sigma}_{22} = -i\hbar^{-1}(\bar{V}_{12}^{ef}\sigma_{21} - \sigma_{12}\bar{V}_{21}^{ef}) - \int_0^t [K_{11}^{11}(\tau)\sigma_{11}(t-\tau) + K_{11}^{22}(\tau)\sigma_{22}(t-\tau)]d\tau \tag{4.11}$$

$$\dot{\sigma}_{12} = \dot{\sigma}_{21}^{*} = -i\omega_{12}\sigma_{12} - i\hbar^{-1}\bar{V}_{12}^{ef}(\sigma_{22} - \sigma_{11}) - \int_0^t [K_{12}^{12}(\tau)\sigma_{12}(t-\tau) + K_{12}^{21}(\tau)\sigma_{21}(t-\tau)]d\tau \tag{4.12}$$

where $\omega_{12} = (J_1 - J_2)/h$.

For the two-state system (with $V_{12} = V_{21}^{*}$; $V_{11} = V_{22} = 0$), according to formula (3.55), the only non-zero matrix elements of $K_{mn}^{m'n'}$ are

$$K_{11}^{11}(\tau) = 2\hbar^{-2}\mathrm{Re}[\langle \tilde{V}_{12} V_{21}(\tau)\rangle e^{-i\omega_{12}\tau}]$$

$$K_{11}^{22}(\tau) = -2\hbar^{-2}\mathrm{Re}[\langle \tilde{V}_{21}(\tau)\tilde{V}_{12}(0)\rangle e^{-i\omega_{12}\tau}]$$

$$K_{12}^{12}(\tau) = \hbar^{-2}\langle \tilde{V}_{21}(0)\tilde{V}_{12}(\tau) + V_{12}(\tau)\tilde{V}_{21}(0)\rangle = K_{21}^{21}(\tau)^{*} \tag{4.13}$$

$$K_{12}^{21}(\tau) = -\hbar^{-2}\langle \tilde{V}_{12}(\tau)\tilde{V}_{12}(0) + \tilde{V}_{12}(0)\tilde{V}_{12}(\tau)\rangle = K_{21}^{12}(\tau)^{*}$$

(Here and further on we omit indices VV and the sign of mean in $K_{mn}^{m'n'}$). In the Markovian limit (3.58), when the equations (3.59) may be used, we get, instead of (4.11) and (4.12), the following equations for the density matrix of the electronic subsystem

$$\dot{\sigma}_{11} = -\dot{\sigma}_{22} = -i\hbar^{-1}\left(\bar{V}_{12}^{ef}\sigma_{21} - \sigma_{12}\bar{V}_{21}^{ef}\right) - \left(w_{12}\sigma_{11} - w_{21}\sigma_{22}\right) \tag{4.14}$$

$$\dot{\sigma}_{12} = \dot{\sigma}_{21}^{*} = -i\omega_{12}\sigma_{12} - i\hbar^{-1}\bar{V}_{12}^{ef}(\sigma_{22} - \sigma_{11}) - \sigma_{12}/T_2 - \sigma_{21}/T_3 \tag{4.15}$$

where according to (3.61) and (4.13) the quantities w_{12}, w_{21}, T_2 and T_3 are defined by the formulae

$$\begin{aligned} w_{12} &= \int_0^\infty K_{11}^{11}(\tau)\,d\tau\,, & w_{21} &= -\int_0^\infty K_{11}^{22}(\tau)\,d\tau \\ T_2^{-1} &= \int_0^\infty K_{12}^{12}(\tau)e^{i\omega_{12}\tau}\,d\tau\,, & T_3^{-1} &= \int_0^\infty K_{12}^{12}(\tau)\,e^{i\omega_{12}\tau}\,d\tau \end{aligned} \tag{4.16}$$

It is worth-while to mention that the equations (4.14) and (4.15), apart from the term with T_3, coincide with the Bloch equations [31] for spin $\frac{1}{2}$ in the external field, $\bar{V}^{ef}$ representing the interaction energy with this field. The term with T_3 is essential in the case of almost degenerate levels J_1 and J_2 when conditions of (3.64) type are not fulfilled. The meaning of the terms with T_3^{-1} in the equation (4.15) is the transformation $\sigma_{12} \rightleftarrows \sigma_{21}$ while in the usual Bloch equations such transformation does not exist. In the case

$$|\omega_{12}| \gg |\bar{V}^{ef}|/\hbar\,,\ w_{ik}\,,\ T_2^{-1}\,,\ T_3^{-1} \tag{4.17}$$

which corresponds to the condition (3.64), the equation (4.14), in accordance with the general result (3.68), obtains the form

$$\dot{\sigma}_{11} = -\dot{\sigma}_{22} = -\left(w_{12}\sigma_{11} - w_{21}\sigma_{22}\right) \tag{4.18}$$

which coincides with the master equation for the two-level system.

From the relations (4.16), (4.8) and (4.3) and after performing averaging of (2.57) type, we obtain the following expressions for the rate coefficients

$$w_{12} = \frac{V^2}{\hbar^2} e^{-F(0)} \int_{-\infty}^{\infty} \left(e^{F(\tau)} - 1\right) e^{-i\omega_{12}\tau} d\tau \tag{4.19}$$

$$w_{21} = \frac{V^2}{\hbar^2} e^{-F(0)} \int_{-\infty}^{\infty} \left(e^{F(\tau)} - 1\right) e^{i\omega_{12}\tau} d\tau \tag{4.20}$$

$$T_2^{-1} = \frac{V^2}{\hbar^2} e^{-F(0)} \int_0^{\infty} \left[\left(e^{F(\tau)} - 1\right) + c.c.\right] e^{i\omega_{12}\tau} d\tau \tag{4.21}$$

$$T_3^{-1} = -\frac{V^2}{\hbar^2} e^{-F(0)} \int_0^{\infty} \left[\left(e^{-F(\tau)} - 1\right) + c.c.\right] e^{i\omega_{21}\tau} d\tau \tag{4.22}$$

and

$$\bar{V}^{ef} = V e^{-\frac{1}{2}F(0)} + (i\hbar)^{-1} \int_0^{\infty} \langle [\tilde{V}(\tau), G(0)] \rangle d\tau \tag{4.23}$$

Here

$$F(\tau) = \sum_k \eta_k^2 \left[\bar{n}_k e^{-i\omega_k\tau} + (n_k + 1) e^{i\omega_k\tau}\right] \tag{4.24}$$

$\bar{n}_k$ - are thermal equilibrium phonon occupation numbers and η_k^2 is defined by (4.4).

At this point it is worthwhile to summarize the analysis of this section.

First, we have applied the general theory of Chapter III to the concrete model system widely used for the description of electron transfer (see Chapter 2.4) and other nonradiative processes. In this model the electron-nuclear system is represented in the Born-Oppenheimer approximation by two intersecting electronic hypersurfaces with a continuum of degrees of freedom plus small perturbation causing transitions between these electronic states. The hypersurfaces are hyperparabolas (2.46) with identical frequencies and different minima.

Second, it has been shown (2.50) that this model may be reduced to the two-level system (4.9) interacting with the phonon bath (4.10). It has been also shown that the equations determining time behavior of the electronic system, generally speaking, do not coincide with master equations even in the Markovian limit. In this case, time evolution of the system is determined by 5 parameters w_{12}, w_{21}, T_2, T_3, $\bar{V}/\hbar^{ef}$ and not only by the transition probabilities as it is in the case of validity of master equations. Below (Section 4.4) we will show the examples when this distinction is important.

Third, the transition probabilities w_{12}, w_{21}, (4.19), (4.20), do not coincide exactly with those widely accepted as rate constants and derived from the Golden Rule formula (see (2.58)). The difference is in the term -1 in the formulae (4.19), (4.20). This term eliminates singularity in these transition probabilities when $\omega_{12} = 0$ [1)] (see (2.59)). But just in this case of zero energy gap (or close to zero), not only transition probabilities but the parameters T_2, T_3 and $\bar{V}^{ef}$ are essential for the description of the system's time evolution.

4.2 Calculation of rate coefficients by saddle-point method

The transition probabilities (4.19), (4.20) are rate coefficients in the master equations (4.18). Now, we will try to examine these quantities in various regions of parameters (such as temperature and reorganization energy (4.6)).

As we know from the general theory (Chapter 1.3) the transition probabilities (per unit time) make sense only for continuous (or quasi-continuous) spectra of energies. More than that, it has been shown that a continuous spectrum of energies is a necessary condition of irreversible motion in general. It means that the sum (4.24) may be presented as Fourier integral

$$F(\tau) = \sum_k \zeta_k^2 \left[\bar{n}_k e^{-i\omega_k \tau} + (\bar{n}_k + 1) e^{i\omega_k \tau} \right] = \int_{-a}^{a} \rho(\omega) e^{-i\omega\tau} d\omega \tag{4.24a}$$

1) The importance of such a term with -1 was stressed by Holstein in his celebrated paper on the polaron motion [39].

Here we have introduced the designations

$$\rho(\omega) = f(\omega)\, \eta^2(\omega)\, n(\omega) \tag{4.25}$$

where f(ω) is the frequency distribution

$$\sum_k \to \int_0^a f(\omega)\, d\omega$$

a - is the maximum frequency of the vibrational spectrum;

$$\eta(\omega) = \eta_k(\omega_k)$$

n(ω) is the average number of phonons with frequency ω

$$n(\omega) = \left[e^{\hbar\omega/k_B T} - 1 \right]^{-1} \qquad (\omega > 0) \tag{4.26}$$

and for negative frequencies we define

$$f(-\omega) = f(\omega) \quad , \quad \eta^2(-\omega) = \eta^2(\omega)$$
$$n(-\omega) = n(\omega) + 1 \tag{4.27}$$

In our new designations the transition probability from electronic state 1 to electronic state 2 (4.19) may be written in the form

$$w_{12} = (V/\hbar)^2 \exp\left[-\int_{-a}^{a} \rho(\omega)\, d\omega \right] \int_{-\infty}^{\infty} d\tau\, e^{-i\omega_{12}\tau} \times$$
$$\times \left[\exp\left(\int_{-a}^{a} \rho(\omega)\, e^{-i\omega\tau}\, d\omega \right) - 1 \right] \tag{4.28}$$

Thus, calculation of the transition probability may be reduced to calculation of the integral

$$I(\lambda) = \int_{-\infty}^{\infty} e^{\psi(\tau)}\, d\tau = \int_{-\infty}^{\infty} e^{F(\tau)}\, e^{i\lambda\tau}\, d\tau \tag{4.29}$$

$$\Psi(\tau) = F(\tau) + i\lambda\tau = \int_{-a}^{a} \rho(\omega) e^{-i\omega\tau} d\omega + i\lambda\tau \tag{4.30}$$

This integral has typical form suitable for calculation by the saddle point method (see, e.g. [40]).

In various limiting cases it may be calculated by this method. According to it, the exponent in (4.29) should contain a large factor. (A bit later we would clarify what it does mean). Then, this factor tending to the infinity, the integral (4.29) has the asymptotic representation

$$I(\lambda) = \sum_{z_0}{}' e^{\Psi(z_0)} \left[2\pi / \left(-\Psi''(z_0) \right) \right]^{1/2} \tag{4.31}$$

where z_0 are all saddle points along the integration contour shifted to the complex region. The saddle points satisfy the equations

$$\Psi'(z_0) = 0 \tag{4.32}$$

The saddle point method is based on expansion of the function $\Psi(t)$ in the vicinity of the saddle point

$$\Psi(z) = \Psi(z_0) + \frac{1}{2} \Psi''(z_0) (z - z_0)^2 \tag{4.33}$$

up to terms of second order in $(z-z_0)$. It means, that in the region where second term in (4.33) gives essential contribution to the integral $I(\lambda)$

$$|z - z_0| \lesssim |\Psi''(z_0)|^{-2}$$

the next terms of the expansion of the function $\Psi(z)$ should be small:

$$|\Psi'''(z_0)(z-z_0)^3| \sim |\Psi'''(z_0)|/|\Psi''(z_0)|^{3/2} \ll 1$$

$$|\Psi^{IV}(z_0)(z-z_0)^4| \sim |\Psi^{IV}(z_0)|/|\Psi''(z_0)|^2 \ll 1$$

(4.34)

Just these conditions determine a large dimensionless factor which is necessary for the application of the saddle point method.

Now we will analyze integral I(λ) (4.29), (4.31) in various limiting cases [41].

In the complex region

$$z_0 = x_0 + iy_0 \tag{4.35}$$

the equation (4.32) takes the form

$$-i\int_{-a}^{a} \omega\rho(\omega) e^{-i\omega x_0} e^{\omega y_0} + i\lambda = 0 \tag{4.36}$$

or

$$\int_{-a}^{a} \omega\rho(\omega) \sin\omega x_0 \, e^{\omega y_0} d\omega = 0$$

$$\int_{-a}^{a} \omega\rho(\omega) \cos\omega x_0 \, e^{\omega y_0} d\omega = \lambda \tag{4.37}$$

We will now consider the case corresponding to small $|y_0|$

$$a|y_0| \ll 1 \tag{4.38}$$

In this approximation we find from (4.36), (4.37) the equations determining x_o, y_o

$$\int_{-a}^{a} \omega \rho(\omega) \sin \omega x_o \, d\omega = 0 \tag{4.39}$$

$$y_o = \frac{\lambda - \int_{-a}^{a} \rho(\omega) \omega \cos \omega x_o \, d\omega}{\int_{-a}^{a} \rho(\omega) \omega^2 \cos \omega x_o \, d\omega} \tag{4.40}$$

Formulae (4.30), (4.31), (4.39) and (4.40) determine the integral (4.29) in the saddle point approximation and the approximation (4.38). Generally speaking, the contributions to this integral give infinite multitude of saddle points. And, apart from exceptional cases, the summation of such infinite sums cannot be performed in an analytical way. That is why it is important to find the conditions under which the main contribution to the integral $I(\lambda)$ (4.29), (4.31) comes from just one stationary point $x_o = 0$ [41]. For this purpose we shall examine the expression for $\Psi(z_o)$ in the approximation (4.38). From (4.30), (4.39), (4.40) it follows that

$$\Psi(z_o) = \int_{-a}^{a} \rho(\omega) \cos \omega x_o \, d\omega - \frac{\left(\lambda - \int_{-a}^{a} \omega \rho(\omega) \cos \omega x_o \, d\omega\right)^2}{2 \int_{-a}^{a} \rho(\omega) \omega^2 \cos \omega x_o \, d\omega}$$

$$- i\lambda x_o - i \int_{-a}^{a} \rho(\omega) \sin \omega x_o \left(1 + \frac{1}{2} \omega y_o^2\right) d\omega \tag{4.41}$$

It is easy to show that in the approximation (4.38) the main contribution to the exponent (4.31) at the saddle point z_o gives first term in (4.41). It means that ratio of a term in (4.31) with $x_o \neq 0$ to that of the one with $x_o = 0$ is proportional to

$$\frac{e^{\Psi(x_o \neq 0)}}{e^{\Psi(x_o = 0)}} \propto e^{\int_{-a}^{a} \rho(\omega)(1 - \cos \omega x_o) \, d\omega}$$

Thus the condition of using only one saddle point with $x_o = 0$ takes the form

$$\int_{-a}^{a} \rho(\omega)\left(1-\cos\omega x_o\right)d\omega \gg 1 \tag{4.42}$$

where x_o is a non-zero root of (4.39). It is easy to check that condition (4.34) of the applicability of the saddle point method (at point $x_o = 0$) may be written in the form

$$\int_{-a}^{a} \rho(\omega)\,d\omega \gg 1 \tag{4.43}$$

which is consistent with (4.42). And at last, the condition of our approximation (4.38) in the explicit form is the following

$$\left|\lambda - \int_{-a}^{a} \rho(\omega)\,\omega\,d\omega\right| \ll \frac{1}{a}\int_{-a}^{a} \rho(\omega)\,\omega^2\,d\omega \tag{4.44}$$

Provided all these conditions are fulfilled, the expression for the transition probability (4.28) obtains the form [see (4.28), (4.29), (4.31), (4.41)]

$$W_{12} = \frac{V^2}{\hbar^2}\left[\frac{2\pi}{\int_{-a}^{a}\rho(\omega)\,\omega^2\,d\omega}\right]^{1/2} \exp\left[-\frac{\left(\omega_{12}-\int_{-a}^{a}\omega\,\rho(\omega)\,d\omega\right)^2}{2\int_{-a}^{a}\rho(\omega)\,\omega^2\,d\omega}\right] \tag{4.45}$$

(For $\Psi''(z_o)$ in the formula (4.31) we got

$$\Psi''(z_o) = -\int_{-a}^{a}\rho(\omega)\,\omega^2\,d\omega).$$

Now we will perform some simple transformations and present (4.45) and condition (4.44) in a bit different form. Using definitions (4.25), (4.27) and (4.4), it is easy to show that

$$-\int_{-a}^{a} \omega \rho(\omega)\, d\omega = E_r/\hbar \tag{4.46}$$

where the reorganization energy E_r is defined by the expression (4.6). Another integral in the formula (4.45) may be presented as

$$\int_{-a}^{a} \rho(\omega)\, \omega^2 d\omega = \int_0^a f(\omega) \xi^2(\omega) \left[2n(\omega)+1\right] \omega^2 d\omega$$
$$\simeq 2k_B T E_r/\hbar^2 \tag{4.47}$$

The latter approximate relation is valid in the high temperature region

$$k_B T \gg \hbar a \tag{4.48}$$

In this region the transition probability (4.45) obtains the form of the Arrhenius law

$$W_{12} = \frac{V^2}{\hbar^2} \left[\frac{\pi \hbar^2}{k_B T E_r}\right]^{1/2} e^{-E_a/k_B T} \tag{4.49}$$

where the activation energy has the form

$$E_a = \frac{(J_1 - J_2 + E_r)^2}{4E_r} \tag{4.50}$$

This essentially the result obtained by Levich [21] from his quantum theoretical treatment of electron transfer.

It is worthwhile mentioning that condition (4.44) with $\lambda = -\omega_{12}$

$$\left|\omega_{21} + \int_{-a}^{a} \rho(\omega)\omega \, d\omega\right| = \hbar^{-1}\left|J_1 - J_2 - E_r\right| \ll \int_{-a}^{a} \rho(\omega)\omega^2 \, d\omega \tag{4.51}$$

may be fulfilled in two cases:

(a) when

$$J_2 - J_1 \simeq E_r \tag{4.52}$$

(b) and when

$$\left|\int_{-a}^{a} \rho(\omega)\omega \, d\omega\right| = \int_{0}^{a} \omega f(\omega)\, \xi^2(\omega)\, d\omega \ll$$

$$\ll \frac{1}{a}\int_{-a}^{a} \rho(\omega)\omega^2 \, d\omega = \frac{1}{a}\int_{0}^{a} f(\omega)\,\xi^2(\omega)\,\omega^2 (2n+1)\, d\omega \tag{4.53}$$

$$J_2 - J_1 \not\simeq E_r$$

The latter relation implies that

$$n(\omega) \gg 1$$

or that the relation (4.48) should be satisfied. Thus the validity of our approximation is limited by high temperatures (4.48), when $J_1 - J_2$ is not close to E_r (i.e. (4.52) is not valid). In this case transition probability w_{12} has an activation-type temperature dependence (4.49). On the other hand the condition (4.51) may be fulfilled at all temperatures provided the difference $J_2 - J_1 - E_r$ is small enough. In this case the expression (4.45) is valid at all temperatures. Transitions of (4.52) type are sometimes called [42] activation-less or barrier-less transitions transitions (depending on the initial state 2 or 1). Of course, in all these cases the validity of the approximation is also limited by the condition (4.42) (or (4.30).

In order to clarify the meaning of the conditions (4.42) and (4.51) which restrict the possibility of using the formulae (4.45), (4.49) we consider two models of frequency distributions. These are the Debye model of acoustic phonons and the Einstein model of optical phonons.

In the Debye model the frequency distribution function is

$$f(\omega) = 3\omega^2/\omega_D^3 \quad , \qquad a = \omega_D \tag{4.54}$$

Of course, this formula has a general meaning (not just in the Debye model) for the frequency distribution of long wave length acoustical phonons. As is known, the number of the phonon states is equal

$$d\tau = V\, dk_x\, dk_y\, dk_z / (2\pi)^3$$

where V is the volume of a system, and $\vec{k}$ are the phonon wavevectors. For acoustical phonons $\vec{k}$ are proportional to frequencies and this implies formula (4.54). In the case when the whole frequency range is essential, from 0 to maximum frequency a, formula (4.54) has no general meaning and may be used only for approximate estimates.

In the long wavelength approximation we may also derive frequency dependence of $\eta^2(\omega)$. Terms of $-n_i\omega_k^2 q_{ik}^0 q_k$ type in the Hamiltonian (4.1) have meaning of change of the electronic energy due to the deformation wave in the medium. This deformation wave in the long wavelength approximation is proportional to the wavevector and thus to the frequency of this wave.*) It means that

$$\omega_k^2 q_{ik}^0 = \alpha_i \omega_k \quad , \qquad q_{ik}^0 = \alpha_i/\omega_k \tag{4.55}$$

where α_i is the coefficient of the proportionality which is dependent on the electronic state i.

Now, from (4.4) and (4.55) we get

$$\eta(\omega) = (\omega/2\hbar)^{1/2}(\alpha_1 - \alpha_2)/\omega = (\alpha_1 - \alpha_2)/(2\hbar\omega)^{1/2}$$

---

*) The force acting on the medium is equal $-\frac{\partial \mathcal{H}}{\partial q}$, and its part connected with change of the electronic energy is $-\frac{\delta \mathcal{H}}{\delta q_k} = -(n_1\omega_k^2 q_{1k}^0 + n_2\omega_k^2 q_{2k}^0)$. This quantity should be proportional to the deformation.

From this relation, from (4.46), (4.25), (4.27) and (4.46) follows the low frequency dependence of $\eta^2(\omega)$

$$\eta^2(\omega) = E_r / \hbar\omega \tag{4.56}$$

In the Debye model approximation we will assume this dependence for the whole frequency range $(0 - \omega_D)$. In this approximation and at high temperatures (4.48) we obtain for $\rho(\omega)$ (4.25) the following expression

$$\rho(\omega) = \frac{3E_r}{\hbar\omega_D} \frac{k_B T}{\hbar\omega_D} \frac{1}{\omega_D} \tag{4.57}$$

Now, the equation for the real part of the saddle point (4.39) takes the form

$$\omega_D x_o \cos \omega_D x_o = \sin \omega_D x_o \tag{4.58}$$

The condition (4.42) of the saddle point method applicability obtains the form

$$6 \frac{E_r}{\hbar\omega_D} \frac{k_B T}{\hbar\omega_D} \left(1 - \frac{\sin \omega_D x_o}{\omega_D x_o}\right) \gg 1 \quad , \quad x_o \neq 0 \tag{4.59}$$

Using the fact that at $x_o \neq 0$,

$$\omega_D x_o \gtrsim 1 \tag{4.60}$$

and using the relation (4.51) we obtain the following applicability conditions of our approximation in the case of the Debye model

$$|J_1 - J_2 - E_r| \ll E_r \frac{k_B T}{\hbar\omega_D} \quad , \quad \frac{E_r}{\hbar\omega_D} \frac{k_B T}{\hbar\omega_D} \gg 1 \tag{4.61}$$

In the Einstein model the vibrational frequencies are restricted to a narrow region near some eigenfrequency ω_o

$$f(\omega) = \begin{cases} 1/2\delta & \text{for } \omega_o - \delta \leq \omega \leq \omega_o + \delta \\ 0 & \text{for the rest } \omega > 0 \end{cases} \tag{4.62}$$

$$\gamma^2(\omega) = \frac{E_r}{\hbar\omega_o} \quad , \quad n(\omega) = \left(e^{\hbar\omega/k_B T} - 1\right)^{-1} \tag{4.63}$$

and we assume that

$$\delta \ll \omega_o \tag{4.64}$$

This model may be suitable for the description of optical phonons. Let us first examine the condition (4.42) and the equation for the real part x_o of the saddle point. For $\delta \to 0$ from the equation (4.39) it follows that

$$\sin \omega_o x_o = 0 \tag{4.65}$$

This equation has roots equal to

$$\omega_o x_o = 2\pi k \quad , \qquad k = 0, \pm 1, \pm 2, \ldots \tag{4.66}$$

It is clear that in this case the condition (4.42) cannot be satisfied. For finite but small δ the root of the equation (4.39) may be written in the form

$$x_o = 2\pi k/\omega_o + \Delta \tag{4.67}$$

where Δ is a small quantity tending to zero together with δ. Using the definition (4.25), (4.27), (4.62 we can rewrite the equation (4.39) in the form

$$\int_{\omega_o-\delta}^{\omega_o+\delta} F(\omega) \sin \omega x_o \, d\omega = 0 \tag{4.69}$$

where

$$F(\omega) = \omega f(\omega) \xi^2(\omega) (2n+1) \tag{4.69}$$

Let us find the first non-zero root of (4.69)

$$X_0 = 2\pi/\omega_0 + \Delta$$

Substituting this expression into (4.68) and performing identical transformations we get

$$\int_{-\delta}^{\delta} F(\omega_0+\xi) \sin\left[\frac{2\pi\xi}{\omega_0} + \Delta(\omega_0+\xi)\right] d\xi = 0$$

Then, again using the smallness of δ (and $\omega_0\Delta$) (4.64) we obtain

$$\omega_0 \Delta \int_{-\delta}^{\delta} F(\omega_0+\xi)\, d\xi = -\frac{6\pi}{3} \delta^2 \frac{F'(\omega_0)\delta}{\omega_0} \tag{4.70}$$

According to (4.62), (4.69) the quantity

$$F'(\omega)\delta$$

is finite when $\delta \to 0$ and does not depend on δ. Thus Δ is proportional to

$$\Delta \propto \frac{\delta^2}{\omega_0^2 \omega_0'} \quad , \quad \omega_0' = \left[\frac{1}{F(\omega_0)} \frac{\partial F(\omega_0)}{\partial \omega_0}\right]^{-1} \tag{4.71}$$

and the condition (4.42) takes the form

$$\frac{E_r}{\hbar\omega_0} (n(\omega_0)+1) \frac{\delta^4}{(\omega_0\omega_0')^2} \gg 1 \tag{4.72}$$

The condition (4.51) may be now written in the form

$$\left| J_1 - J_2 - E_r \right| \ll \frac{E_r}{\hbar\omega_0}\left(n(\omega_0) + 1 \right) \tag{4.73}$$

Of course, in the high temperature region (4.48) the expression for the rate constant (4.49) has the same form for all frequency distributions, including (4.54) and (4.62).

Now, we would examine another limiting case in which the saddle point method may be applied. This case corresponds to large energy gaps

$$\left| J_1 - J_2 \right| \gg \hbar a \tag{4.74}$$

and to large parameter $a|y_0|$, appearing on the saddle point method (4.35)

$$a\,|y_0| \gg 1 \tag{4.75}$$

A bit later we will decode the meaning of this condition. It enables us to calculate the integral (4.36) up to terms of arbitratry order of $1/a|y_0|$. Performing the integration by parts, one obtains from (4.36), (4.75) the equations for saddle points x_0, y_0. These equations are accurate up to terms of order $(ay_0)^{-1}$

$$\lambda = \frac{e^{a|y_0|}}{a|y_0|}\, \rho\left(\frac{\lambda}{|\lambda|} a \right) a^2 \cos a x_0 \tag{4.76}$$

$$\operatorname{tg} a x_0 = \frac{a x_0}{a|y_0|} \tag{4.77}$$

Solving approximately these equations and using again condition (4.75) we get

$$a|y_0| = \ln\left(\frac{|\lambda|}{\rho\left(\frac{\lambda}{|\lambda|} a\right) a^2 \cos a x_0} \right) + \ln\ln\left(\frac{|\lambda|}{\rho\left(\frac{\lambda}{|\lambda|} a\right) a^2 \cos a x_0} \right) \tag{4.78}$$

$$a x_0 = 2\pi k + 2\pi k / |y_0| a \tag{4.79}$$

We see that in this case many saddle points give the contributions to the integral (4.29), (4.31). Calculating the exponent $\Psi(z_o)$ and $\Psi''(z_o)$ we get (neglecting terms of order $1/a^2|y_o|^2$)

$$\Psi(z_o) = -|\lambda y_o| + \frac{|\lambda|}{a} + i\frac{\lambda}{a}2\pi k + i\frac{\lambda}{a}\frac{2\pi k}{|y_o|a} \tag{4.80}$$

$$\Psi''(z_o) = -|\lambda|a \tag{4.81}$$

First, we will examine the contribution of zero saddle point

$$x_o = 0 \quad , \quad k = 0$$

$$a|y_o| = \ln\frac{|\lambda|}{\rho\left(\frac{\lambda}{|\lambda|}a\right)a^2} + \ln\ln\frac{|\lambda|}{\rho\left(\frac{\lambda}{|\lambda|}a\right)a^2} \tag{4.82}$$

Using (4.31), (4.80), (4.81), (4.82) we get an expression for the integral $I(\lambda)$

$$I(\lambda) = \left(\frac{2\pi}{|\lambda|a}\right)^{1/2}\exp\left\{-\frac{|\lambda|}{a}\left[\ln\frac{|\lambda|}{\rho\left(\frac{\lambda}{|\lambda|}a\right)a^2} + \ln\ln\frac{|\lambda|}{\rho\left(\frac{\lambda}{|\lambda|}a\right)a^2} - 1\right]\right\} \tag{4.83}$$

Now, we will clarify the validity conditions of this expression. As we know the saddle point method may be applied provided conditions of (4.34) type are fulfilled. Calculating higher derivatives of $\Psi(z_o)$ one can verify that conditions (4.34) are satisfied if

$$|\lambda|/a \gg 1 \tag{4.84}$$

This coincides with condition (4.74) which we have assumed to be fulfilled. Another condition (4.75), which we used in calculating the integrals, may be written now according to (4.82) in the form

$$\ln\frac{|\lambda|}{\rho\left(\frac{\lambda}{|\lambda|}a\right)a^2} = \ln\gamma \gg 1 \tag{4.85}$$

Before proceeding further, an important comment should be made. We have approximately calculated $a|y_0|$ (4.82) assuming the condition (4.85) is fulfilled. Solving the equation (4.76) for $|y_0|$ in the higher approximation, it is easy to verify that the correction of the next approximation has the form $\ln\ln\ln\gamma$. This quantity is much smaller than $\ln\gamma$ and $\ln\ln\gamma$. But we should remember that this quantity stands in the exponent $\Psi(z_0)$ where it is multiplied by $|\lambda|/a$. Thus it corrects $\Psi(z_0)$ by the order of magnitude

$$\frac{|\lambda|}{a}\ln\ln\ln\gamma \tag{4.86}$$

Because of the assumption (4.85) and (4.74) (which means $\frac{|\lambda|}{a} \gg 1$ and which is necessary for the application of the saddle point method), the correction (4.86) is greater than unity. It should also be mentioned that the above neglect of terms of order $1/a^2|y_0|^2$ for determining the saddle points leads to the error in the exponent index of the order

$$\frac{|\lambda|}{a}\frac{1}{|y_0|a} \sim \frac{|\lambda|}{a}\frac{1}{\ln(|\lambda|/a)} \tag{4.87}$$

which is greater than unity as well. Thus, in our approximation the integral (or more exactly - the contribution to it of the saddle point with $x_0 = 0$) has been calculated with so-called logarithmic accuracy. It means that the exponent index $\Psi(z_0)$ (or $\ln I(\lambda)$) has been expanded into a series where each next term is much smaller than the preceding, and only the first leading terms were preserved. But since the corrections in the exponent index are larger than unity, the corrections to a preexponent factor are not small. The same refers to other saddle points with $x_0 \neq 0$. All of them have the same order of magnitude as that of the one with $x_0 = 0$. But since they have been calculated only with logarithmic accuracy the summation over all saddle-points z_0 makes no sense. It means that the saddle point method is not appropriate for the calculation of the integral $I(\lambda)$ in the limit (4.74), (4.75). In this case an alternative method of the integral equation [42] may be used.

4.3 Integral equation for nonradiative transition rates

An integral equation may be derived for the integral

$$J(\lambda) = \int_{-\infty}^{\infty}\left[e^{\int_{-a}^{a}\rho(\omega)e^{-i\omega\tau}d\omega} - 1\right]e^{i\lambda\tau}\,d\tau \tag{4.88}$$

determining the transition rate (4.28). Performing integration by parts in the expression (4.88) and taking into account that in any physical situation (for continuous spectrum of energies, see chapter 1.3)

$$\lim_{t \to \pm\infty} \int_{-a}^{a} \rho(\omega) e^{-i\omega t} d\omega = 0$$

we get

$$J(\lambda) = 2\pi\rho(\lambda) + \frac{1}{\lambda} \int_{-a}^{a} \rho(\omega)\, \omega\, J(\lambda-\omega)\, d\omega \tag{4.89}$$

This relation constitutes a new integral equation for the nonradiative transition rates [43]*). According to (4.28) the transition rate may be expressed through this quantity $J(\lambda)$ as follows

$$w_{12} = \frac{V^2}{\hbar^2} e^{-\int_{-a}^{a} \rho(\omega) d\omega} J(\omega_{21}) \tag{4.90}$$

The expression (4.28), and thus (4.90), have been derived for the specific model of the electron-nuclear system. In this model the potential hypersurfaces are hyperparabolas (2.46) with identical frequencies (curvatures) and displaced minima, and the perturbation energy does not depend on normal coordinates q_k. In a general case when the energy hypersurfaces are characterized by a different set of frequencies, or the perturbation energy depends on q_k or on $\partial/\partial q_k$ [44, 45], the expression for w_{12} is more complicated and the integral determining it contains a pre-exponential factor $F(\tau)$

$$I(\lambda) = \frac{1}{2\pi} \int_{-\infty}^{\infty} e^{i\lambda\tau} F(\tau) \exp\left[\int_{-a}^{a} \rho(\omega) e^{i\omega\tau} d\omega\right] d\tau$$

---

*) The equation of the paper [43] contains $\delta(\lambda)$. This is connected with the fact that it has been derived for the quantity $I(\lambda) = \frac{1}{2\pi} \int_{\infty}^{\infty} e^{i\lambda t} \exp[\int_{-a}^{a} \rho(\omega) e^{-i\omega t} d\omega] dt$ which contains the singularity at $\lambda = 0$.

But also, in this case, the quantity $I(\lambda)$ may be represented in the form (4.88) with a renormalized $\rho(\omega) \to \rho(\omega) + \rho_1(\omega)$, and $\ln F(t) = \int \rho_1(\omega) e^{-i\omega t} d\omega$. Thus (4.88)-(4.90) may be considered as a rather general equation for derivation of the transition rate expressions.

Using our integral equation (4.89) we can check the validity of expressions found (section 4.2) by the saddle point method. In the case of high reorganization energies and high temperatures (4.61), (4.43), (4.44) (which correspond to strong electron-phonon coupling), the integral $J(\lambda)$ may be written in the form (see (4.45), (4.90))

$$J(\lambda) = (2\pi)^{\frac{1}{2}} \left(\int_{-a}^{a} \rho(\omega) \omega^2 d\omega \right)^{-\frac{1}{2}} \exp\left(\int_{-a}^{a} \rho(\omega) d\omega \right) \times$$

$$\times \exp\left[- \frac{\left(\lambda - \int_{-a}^{a} \rho(\omega) \omega d\omega \right)^2}{2 \int_{-a}^{a} \rho(\omega) \omega^2 d\omega} \right] \qquad (4.91)$$

Substituting this expression for the integral equation (4.89) one can verify [43] that it satisfies this equation provided conditions (4.43), (4.44) are fulfilled.

A simple derivation of the functional dependence (4.91) on λ may be performed directly form the integral equation (4.89). For this purpose one should expand $J(\lambda-\omega)$ in the integrand up to first order terms ω

$$J(\lambda - \omega) = J(\lambda) - J'(\lambda)\, \omega$$

and to substitute for the r.h.s. of (4.89). Neglecting the term $2\pi\rho(\lambda)$ in (4.89), we get the differential equation

$$J'(\lambda) \int_{-a}^{a} \rho(\omega) \omega^2 d\omega + J(\lambda) \left(\lambda - \int_{-a}^{a} \rho(\omega) \omega d\omega \right) = 0 \qquad (4.92)$$

The solution of this equation has λ-dependence (4.91). A condition of derivation (4.92) is

$$a\, J'(\lambda) / J(\lambda) \ll 1$$

which coincides with (4.44). The neglect of $2\pi\rho(\lambda)$ in the r.h.s. of (4.89) corresponds to the condition (4.43).

In the case of large energy gaps (4.74) and weak coupling (4.75), (4.85), the situation is more complicated. In a previous section the solution for $I(\lambda)$ was found (4.83) taking into account only one saddle point $x_o = 0$. This solution is valid with the logarithmic accuracy. One can argue that summation over all stationary points would not change the basic functional dependence form of the rate and would influence only the pre-exponential numerical coefficient k. Equation (4.89) gives us the possibility to check such considerations. Thus we would suppose that both the quantity $J(\lambda)$ and the transition probability may be written in the form*)

$$W_{12} \propto J(\lambda) = k \exp\left\{ -\frac{|\lambda|}{a}\left[\ln\left(\frac{|\lambda|}{\rho(\frac{\lambda}{|\lambda|}a)a^2}\right) + \ln\ln\left(\frac{|\lambda|}{\rho(a\lambda/|\lambda|)a^2}\right) - 1\right]\right\}, \quad \lambda = \omega_{21} \tag{4.93}$$

Substituting this expression into (4.89) and using conditions (4.84), (4.85)

$$|\lambda| \gg a, \quad \ln \frac{|\lambda|}{\rho(a\lambda/|\lambda|)a^2} \gg 1 \tag{4.94}$$

we obtain

$$J(\lambda) \approx \frac{J(\lambda)}{\lambda}\int_{-a}^{a} \exp\left\{ \frac{\omega}{a}\frac{|\lambda|}{\lambda}\left[\ln\left(\frac{|\lambda|}{\rho(a|\lambda|/\lambda)a^2}\right) + \ln\ln\left(\frac{|\lambda|}{\rho(a|\lambda|/\lambda)a^2}\right)\right]\right\}\rho(\omega)\,d\omega$$

Taking this integral by part and using conditions (4.94) we get

$$J(\lambda) \simeq J(\lambda) + O\left[\frac{1}{\ln\left(\frac{|\lambda|}{\rho(a\lambda/|\lambda|)a^2}\right)}\right]$$

*) Since $|\lambda| \gg a$, there is no difference between $I(\lambda)$ and $J(\lambda)$ (this difference is connected with inclusion - 1 in the r.h.s. of (4.88))

Thus the integral equation (4.89) is satisfied by the expression (4.93) provided conditions (4.94) are satisfied. The expression (4.93) may be easily generalized to the case when $\rho(a\frac{|\lambda|}{\lambda})$ and its first n-1 derivatives vanish

$$\rho(a|\lambda|/\lambda) = \rho'(a|\lambda|/\lambda) = \dots = \rho^{(n-1)}(a|\lambda|/\lambda) = 0$$

In this case the integral equation is satisfied by

$$J(\lambda) = k \exp\left\{-\frac{|\lambda|}{a}\left[\ln\left(\frac{|\lambda|(-1)^n}{\rho^{(n)}(a|\lambda|/\lambda)\,a^{n+2}}\right) + (n+1)\ln\ln\left(\frac{|\lambda|(-1)^n}{\rho^{(n)}(a|\lambda|/\lambda)\,a^{n+2}}\right) - 1\right]\right\} \tag{4.95}$$

provided

$$|\lambda|/a \gg 1, \quad \ln\left(|\lambda|(-1)^n/\rho^{(n)}(a|\lambda|/\lambda)\,a^{n+2}\right) \gg 1 \tag{4.96}$$

Now, we will illustrate the dependence (4.93) of the transition rate w_{12} on the energy gap

$$\lambda = (J_2 - J_1)/\hbar$$

in the Debye and Einstein models.

Thus, we consider the transition from the state with energy J_1 to state with energy J_2.

In the Debye model (4.54), (4.56) we get for w_{12} [41]

$$w_{12} = k \exp\left\{-\frac{|J_2-J_1|}{\hbar\omega_D}\left[\ln\left(\frac{|J_2-J_1|}{3E_r(n(\omega_D)+\frac{1}{2}\mp\frac{1}{2})}\right) + \ln\ln\left(\frac{|J_2-J_1|}{3E_r(n(\omega_D)+\frac{1}{2}\mp\frac{1}{2})}\right) - 1\right]\right\} \tag{4.97}$$

The conditions of the applicability of this expression are

$$|J_2-J_1| \gg \hbar\omega_D, \quad |J_2-J_1| \gg 3E_r\left(n(\omega_D)+\frac{1}{2}\mp\frac{1}{2}\right) \tag{4.98}$$

where "-" and "+" signs correspond to $J_2 > J_1$ and to $J_2 < J_1$ respectively (for the transition $1 \to 2$).

For the Einstein model (4.62) the applicability conditions transform to the form

$$|J_2 - J_1| \gg \hbar\omega_0\,, \quad |J_2 - J_1| \gg E_r(\omega_0/\delta)\left(n(\omega_0) \mp \tfrac{1}{2} + \tfrac{1}{2}\right) \tag{4.99}$$

and the transition rate is given by [41]

$$W_{12} = k \exp\left\{-\frac{|J_1 - J_2|}{\hbar\omega_0}\left[\ln\left(\frac{|J_1 - J_2|}{E_r}\frac{\delta}{\omega_0\left(n(\omega_0) + \frac{1}{2} \pm \frac{1}{2}\right)}\right) + \ln\ln\left(\frac{|J_2 - J_1|\,\delta}{E_r\omega_0\left(n(\omega_0) + \frac{1}{2} \pm \frac{1}{2}\right)}\right) - 1\right]\right\} \tag{4.100}$$

It is clear that these expressions mainly coincide with the energy gap law [46]. It is easy to show the origin of the dependence expressed by the formula (4.93). When conditions of (4.98), (4.99) type are satisfied the main contribution to the transition rate (4.28) comes from the first non-vanishing term in the expansion of the exponent in (4.28) allowed by the conservation laws, i.e. for

$$|\omega_{21}| \simeq l\omega_D \quad \text{or} \quad l\omega_0$$

This term is proportional to

$$\begin{aligned} W_{12} &\propto \int_{-\infty}^{\infty} d\tau\, e^{i\omega_{21}\tau}\left(\int_{-a}^{a} \rho(\omega)\, e^{-i\omega\tau}\, d\omega\right)^l / l! \\ &\propto \frac{\alpha^l}{l!} \simeq \alpha^l l^{-l} e^l \sqrt{2\pi l} = \sqrt{2\pi l}\, e^{-l(\ln(l/\alpha) - 1)} \\ &\propto \exp\left\{-\frac{|\omega_{21}|}{\omega_D}\left[\ln\left(\frac{|\omega_{21}|}{\omega_D \alpha}\right) - 1\right]\right\} \end{aligned} \tag{4.101}$$

where α is the constant proportional to E_r.

The dependence (4.101) mainly coincides with that of (4.93), (4.97) and (4.100). It means that in the approximation (4.94), (4.98), (4.49) the main contribution to the dependence of the transition rate on the energy gap $J_1 - J_2$ comes from the first non-vanishing ℓ-phonon transition between states J_1 and J_2 allowed by the conservation law

$$\ell \hbar\omega_D \simeq |J_1 - J_2| \quad \text{or} \quad \ell \hbar\omega_0 \simeq |J_1 - J_2| \tag{4.102}$$

With the same logarithmic accuracy, which we are using here, one can neglect terms with $\ln\alpha$, and get

$$w_{12} \propto \exp\left[-\frac{|J_1 - J_2|}{\hbar\omega_{max}} \ln\left(\frac{J_1 - J_2}{\hbar\omega_{max}}\right)\right] \tag{4.102}$$

Thus, this kind of dependence (which is the essence of the energy gap law [46]) may be obtained from quite general considerations and does not depend on the specific model.

4.4 Symmetric case of zero energy gap

Temperature dependence

We continue to consider the time evolution of the system described by the Hamiltonian (4.8). As it has been said, this Hamiltonian may describe electron transfer in condensed media. We have derived the equations of motion for the subsystem (4.14), (4.15). In the case (4.17) of large enough energy gaps $J_1 - J_2$ between electronic states, these equations may be reduced to the master equations (4.18). In the preceding sections 4.2, 4.3, we have examined the transition rates w_{12}, w_{21}, which are kinetic coefficients in the equations (4.18).

Now, we are going to analyze the case of zero (or small) energy gap

$$\omega_{12} = 0 \quad , \quad J_1 = J_2 \tag{4.103}$$

which corresponds to the case of two symmetric electronic hypersurfaces.

In this case the equation of motion for the density matrix σ_{ik} of the electronic subsystem obtains the form (see (4.14), (4.15))

$$\dot{\sigma}_{11} = -\dot{\sigma}_{22} = -i\hbar^{-1}\bar{V}\left(\sigma_{21} - \sigma_{12}\right) - w\left(\sigma_{11} - \sigma_{22}\right) \tag{4.104}$$

$$\dot{\sigma}_{12} = \dot{\sigma}_{21}^{*} = -i\hbar^{-1}\bar{V}\left(\sigma_{22} - \sigma_{11}\right) - w\sigma_{12} - \sigma_{21}/T_3 \tag{4.105}$$

where (see (4.19))

$$w_{12} = w_{21} = T_2^{-1} = w = \frac{|V|^2}{\hbar^2} \exp\left[-\int_{-a}^{a} \rho(\omega)\, d\omega\right] \times$$

$$\times \int_{-\infty}^{\infty} \left[\exp\left(\int_{-a}^{a} \rho(\omega) e^{-i\omega t} d\omega\right) - 1\right] dt \tag{4.106}$$

$$T_3^{-1} = -\frac{V^2}{\hbar^2} \exp\left(-\int_{-a}^{a} \rho(\omega) d\omega\right) \int_{-\infty}^{\infty} \left[\exp\left(-\int_{-a}^{a} \rho(\omega) e^{-i\omega t} d\omega\right) - 1\right] dt \tag{4.107}$$

$$\overline{V} = V \exp\left(-\frac{1}{2} \int_{-a}^{a} \rho(\omega)\, d\omega\right) \tag{4.108}$$

Here $\rho(\omega)$ is defined by relations (4.24)-(4.27), and we have assumed for simplicity

$$V^{ef} = V$$

i.e. we have omitted the second term in the r.h.s. of (3.60). Of course, the equations (4.104), (4.105) have nothing to do with master equations and describe more a complicated (but still irreversible) motion. We shall analyze these equations in various temperature regions, starting from low temperatures.*)

We begin with the expansion of the exponent in the integrand (4.106)

$$w = \frac{2\pi}{\hbar} V^2 \exp\left(-\int_{-a}^{a} \rho(\omega) d\omega\right) \left[\rho(0) + \frac{1}{2!} \int_{-a}^{a} \rho(\omega_1) \rho(-\omega_1) d\omega_1 + \right.$$

$$\left. + \ldots + \frac{1}{k!} \int_{-a}^{a} d\omega_1 \int_{-a}^{a} d\omega_2 \ldots \int_{-a}^{a} d\omega_{k-1}\, \rho(\omega_1) \ldots \rho(\omega_{k-1}) + \ldots\right] \tag{4.109}$$

---

*) In the next chapter it will be shown that at absolute zero and at very low temperatures the equations (4.104), (4.105) and coefficients (4.106), (4.107) should be essentially modified.

Our aim is to find out under what conditions all the integrals in the r.h.s. are small in comparison with $\rho(0)$. First, let us consider the case of low temperatures

$$k_B T/\hbar \ll a \qquad (4.110)$$

It is obvious that in this case, only low-frequency acoustic phonons give non-vanishing contributions to the integrals. According to (4.57) in the low frequency approximation

$$\rho(\omega) \simeq \frac{3E_r}{\hbar\omega_D} \frac{k_B T}{\hbar\omega_D} \frac{1}{\omega_D} \qquad (4.111)$$

The validity of this expression is not limited by the Debye model approximation, though we are using the same designations as in the Debye model. The difference between a (the maximum frequency of the spectrum) and ω_D in (4.57) is that the latter is the characteristic of low frequency behavior of $\rho(\omega)$ and $f(\omega)$ and coincides with a only in the Debye model. In the case when (4.110) is not satisfied and the whole frequency region is essential,

$$0 \leq \omega \leq a$$

we will use the Debye model. For example, to calculate the integral $\int_{-a}^{a} \rho(\omega) d\omega$ we use (4.54) at all frequencies.*)

In the low temperature region (4.110) this integral is equal to

$$\int_{-a}^{a} \rho(\omega)\, d\omega \simeq \frac{3}{2} \frac{E_r}{\hbar\omega_D} \qquad k_B T \ll \hbar\omega_D \qquad (4.112)$$

and at high temperatures

$$\int_{-a}^{a} \rho(\omega)\, d\omega \simeq 6 \frac{E_r}{\hbar\omega_D} \frac{k_B T}{\hbar\omega_D} \qquad (4.113)$$

---

*) It means, particularly, that even at low temperatures the optical phonons may give essential contribution to this integral.

It is easy to show that in the low temperature region (4.110) each succeeding term of the expansion (4.109) is smaller than the preceding one by the factor $(E_r/h\omega_D)(k_BT/h\omega_D)^2$. Therefore, the approximate expression for the transition rate (4.106)

$$W = \frac{2\pi}{\hbar} V^2 \exp\left(-\int_{-a}^{a} \rho(\omega)\, d\omega\right) \rho(0) = \tag{4.114}$$
$$= 2\pi \hbar^{-2} V^2 \exp\left(-\int_{-a}^{a} \rho(\omega)\, d\omega\right) (3/\omega_D)(E_r/\hbar\omega_D)(k_BT/\hbar\omega_D)$$

is valid provided the conditions

$$(E_r/\hbar\omega_D)(k_BT/\hbar\omega_D)^2 \ll 1 \;, \quad k_BT/\hbar\omega_D \ll 1 \tag{4.115}$$

are satisfied. Examining the expansion in the r.h.s. of (4.109) when

$$k_BT/\hbar\omega_D \sim k_BT/\hbar\omega_a \gg 1 \tag{4.116}$$

it is easy to show that the expression (4.117) is valid provided

$$(E_r/\hbar\omega_D)(k_BT/\hbar\omega_D) \ll 1 \tag{4.117}$$

Both these conditions may be satisfied simultaneously provided

$$E_r/\hbar\omega_D \ll 1 \tag{4.118}$$

On the other hand, the conditions (4.115) do not necessarily imply smallness of this ratio $E_r/h\omega_D$, if k_BT is sufficiently small.

In all these cases the quantity T_3^{-1} may be calculated in the same way with the result

$$T_3^{-1} = w \tag{4.119}$$

Now, the equations determining time evolution of the electronic density matrix follow from (4.104), (4.105), (4.106) and (4.119). Introducing the designations

$$r = \sigma_{11} - \sigma_{22} \ , \ \rho = \sigma_{21} - \sigma_{12} \ , \ \sigma = \sigma_{12} + \sigma_{21} \tag{4.120}$$

we get

$$\dot{r} = -i\Omega_0 \rho - 2wr \ , \ \dot{\sigma} = -2w\sigma \ , \ \dot{\rho} = -i\Omega_0 r \tag{4.121}$$

and

$$\ddot{r} + \Omega_0^2 r + 2w\dot{r} = 0 \tag{4.122}$$

where (see also (2.17))

$$\Omega_0 = 2\bar{V}/\hbar = (2V/\hbar)\exp\left(-\int_{-a}^{a}\rho(\omega)\,d\omega\right) = \begin{cases} (2V/\hbar)\exp\left(-\frac{3}{4}E_r/\hbar\omega_D\right), & k_B T < \hbar\omega_D \\ (2V/\hbar)\exp\left[-3(E_r/\hbar\omega_D)(k_B T/\hbar\omega_D)\right], & k_B T > \hbar\omega_D \end{cases} \tag{4.123}$$

Thus, the time evolution of the population difference $r = \sigma_{11} - \sigma_{22}$ is determined in this case by the equation of decaying harmonic oscillator with the frequency Ω_0 (4.123) and decay constant (4.114). It is worthwhile to estimate the ratio of these two quantities

$$w/\Omega_0 \sim (E_r/\hbar\omega_D)(k_B T/\hbar\omega_D)^2 (V/k_B T)\exp\left(-\frac{1}{2}\int_{-a}^{a}\rho(\omega)\,d\omega\right) \tag{4.124}$$

This quantity is small in all cases (4.115)-(4.117) provided $\bar{V}$ is not much larger than $k_B T$ or $\hbar\omega_D$. (The case when $\bar{V} >> k_B T$ will be considered in the next chapter).

Now, we start considering another temperature region

$$\left(E_r/\hbar a\right)\left(k_B T/\hbar a\right)^2 \gg 1 \qquad (4.125)$$

In this case we may use the saddle point method for calculating integrals appearing on (4.106), (4.107).*) We will slightly generalize the problem and consider also small but non-zero energy gaps. In the case (4.125) one can neglect unity in (4.106) and examine the integral

$$I(\lambda) = \int_{-\infty}^{\infty} e^{i\lambda t} \exp\left[\int_{-a}^{a} \rho(\omega) e^{-i\omega t} d\omega\right] dt \equiv$$

$$\equiv \int_{-\infty}^{\infty} \exp\left[\Psi(t)\right] dt \qquad (4.126)$$

The equation for the saddle point (4.32) may be written in the form

$$\Psi'(z_0) = i\lambda - i\int_{-a}^{a} \rho(\omega)\, \omega\, e^{-i\omega z_0} d\omega = 0$$

Utilizing the definitions (4.25)-(4.27) we get

$$\lambda = \int_0^a \omega f(\omega) \xi^2(\omega) \left[n e^{-i\omega z_0} - (n+1) e^{i\omega z_0}\right] d\omega \qquad (4.127)$$

Now, we will assume that the phonon occupation numbers correspond to the thermal equilibrium

$$(n+1)/n = \exp\left(\hbar\omega/k_B T\right) \qquad (4.128)$$

and look for a solution of the equation (4.127) in the form

$$z_0 = i\hbar/2k_B T + iy_0 + x_0 \qquad (4.129)$$

*) Holstein [39] calculated such integrals by the saddle point method considering the problem of small polaron motion.

Performing identical transformations and taking into account (4.128) we get

$$\lambda = \int_0^a \omega f(\omega) \zeta^2(\omega) n(\omega) e^{\hbar\omega/2k_BT} \left[e^{\omega y_0} e^{-i\omega x_0} - e^{-\omega y_0} e^{i\omega x_0} \right] = 0$$

or

$$\lambda = \int_0^a \omega f(\omega) \zeta^2(\omega) n e^{\hbar\omega/2k_BT} \left(e^{\omega y_0} - e^{-\omega y_0} \right) \cos\omega x_0 ;$$

$$\int_0^a \omega f(\omega) \zeta^2(\omega) n e^{\hbar\omega/2k_BT} \left(e^{\omega y_0} + e^{-\omega y_0} \right) \sin \omega x_0 = 0 \qquad (4.130)$$

Assuming that λ (proportional to the energy gap) is small, we get

$$a|y_0| << 1$$

and

$$a y_0 = a\lambda/2 \int_0^a \omega^2 f(\omega) \zeta^2(\omega) n(\omega) e^{\hbar\omega/2k_BT} \cos \omega x_0 \, d\omega \qquad (4.131)$$

$$\int_0^a \omega f(\omega) \zeta^2(\omega) n(\omega) e^{\hbar\omega/2k_BT} \sin \omega x_0 = 0 \qquad (4.132)$$

These equations determine all the saddle points in the approximation of small energy gaps, i.e. when the r.h.s. of (4.131) is small.

Performing the same analysis as in Section 4.2 (the relation (4.42)) we get the following condition of the applicability of the saddle point method (with only one saddle point $x_0 = 0$)

$$\int_0^a f(\omega) \zeta^2(\omega) e^{\hbar\omega/2k_BT} n(\omega) \left(1 - \cos \omega x_0\right) d\omega >> 1 , \quad x_0 \neq 0 \qquad (4.133)$$

This condition is satisfied if $f(\omega)$, $\eta^2(\omega)$ are smooth enough (do not contain δ-function-like singularities) and if the relation (4.125) is satisfied. Another condition which we have used is connected with the smallness of the energy gap ($a|y_0| \ll 1$) and has the form

$$|J_1 - J_2| \ll \frac{\hbar}{a}\int_0^a f(\omega)\,\omega^2 \zeta^2(\omega)\, n(\omega)\, \exp(\hbar\omega/2k_BT)\, d\omega \tag{4.134}$$

Provided these conditions (4.125), (4.133), (4.134) are fulfilled, the expression for the transition rate w_{12} takes the form

$$\begin{aligned} w_{12} = {} & (V^2/\hbar^2) \exp\left(-\int_{-a}^{a} \rho(\omega)\, d\omega\right) \times (2\pi)^{\frac{1}{2}} \times \\ & \times \left(2\int_0^a f(\omega)\zeta^2(\omega)\,\omega^2 n(\omega) \exp(\hbar\omega/2k_BT)\, d\omega\right)^{-\frac{1}{2}} \times \\ & \times \exp\Big[2\int_0^a f(\omega)\zeta^2(\omega)\, e^{\hbar\omega/2k_BT} n(\omega)\, d\omega - \\ & - (J_2 - J_1)^2/4\hbar^2 \int_0^a \omega^2 f(\omega)\zeta^2(\omega)\, n(\omega)\, e^{\hbar\omega/2k_BT}\, d\omega\Big]\, e^{-(J_2-J_1)/2k_BT} \end{aligned} \tag{4.135}$$

In case of zero energy gap

$$J_1 - J_2 = 0$$

this expression may be written in the form

$$\begin{aligned} w = {} & (V^2/\hbar^2) \exp\left(-\int_{-a}^{a} \rho(\omega)\, d\omega\right) (2\pi)^{\frac{1}{2}} \times \\ & \times \left[\int_{-a}^{a} \rho(\omega)\,\omega^2 e^{\hbar\omega/2k_BT}\right]^{-\frac{1}{2}} \exp\left[\int_{-a}^{a} \rho(\omega)\, e^{\hbar\omega/2k_BT}\, d\omega\right] \end{aligned} \tag{4.136}$$

It is also easy to see that in the same approximation (4.125) the r.h.s. of (4.107) is negligibly small in comparison with (4.106), i.e.

$$T_3^{-1} \ll w \tag{4.137}$$

In the case when

$$k_B T > \hbar a \tag{4.138}$$

it is easy to show that (4.136) reduces to the formula (4.49) obtained earlier

$$w = \left(|V|^2/\hbar^2\right)\left(\pi\hbar^2/k_B T E_r\right)^{1/2} \exp\left(-E_r/4k_B T\right) \tag{4.139}$$

In the case (4.137), the equations for the quantities r, ρ, σ (4.120) obtain the form

$$\dot{r} = -i\Omega_0\rho - 2wr \,, \quad \dot{\sigma} = -w\sigma \,, \quad \dot{\rho} = -i\Omega_0 r - w\rho \tag{4.140}$$

and

$$\ddot{r} + \left(\Omega_0^2 + 2w^2\right) r + 3w\dot{r} = 0 \tag{4.141}$$

One can estimate that for

$$k_B T < \hbar a \,, \quad \left(E_r/\hbar a\right)\left(k_B T/\hbar a\right)^2 \gg 1$$

the frequency of the quantum beats Ω_0 is greater than w

$$\Omega_0 \gg w \tag{4.142}$$

On the other hand at high temperatures (4.138)

$$\Omega_0 \ll w \tag{4.143}$$

and the difference of the populations would decay without quantum beats according to the law ($t \gg w^{-1}$)

$$r = r_0 e^{-wt}$$

The same dependence would follow from the master equation for r

$$\dot{r} = -wr \tag{4.144}$$

It means that in the temperature region (4.138), (4.125) we can neglect quantum beats and utilize usual master equation approach.

V. EQUATIONS OF MOTION AT ZERO AND LOW TEMPERATURE RANGE

5.1 Equations of motion in the case when the interaction energy V is not a small quantity

In the preceding Chapter we have analyzed the equations of motion and their coefficients - transition rates - using a comparatively simple model presented by the Hamiltonian (4.1) or (4.8). These Hamiltonians were used in the literature for description of various radiationless processes. In Chapter II the electron transfer and energy transfer were presented as examples of the processes which may be modeled by these Hamiltonians. The whole consideration of Chapter IV was based on the assumption that the interaction energy V in (4.8) is a small quantity. The exact meaning of this assumption will be analyzed in the next Section 4.2.

However, some preliminary remarks should be put here. It is clear from the above that quantity characterizing interaction with the medium is not V itself, but the quantity $\bar{V}$, averaged over the phonon bath (see (4.23)), i.e. multiplied by the Debye-Waller factor

$$\bar{V} = V \exp\left[-\frac{1}{2}\sum_k \gamma_k^2 (2\bar{n}_k + 1)\right] = V \exp\left[-\frac{1}{2}\int_{-a}^{a} \rho(\omega)\, d\omega\right] \tag{5.1}$$

A-priori, there exist several dimensionless parameters containing $\bar{V}$:

$$\xi_1 = \bar{V}/|J_1 - J_2|\,,\quad \xi_2 = \bar{V}/k_B T\,,\quad \xi_3 = \bar{V}/\hbar\omega^* \simeq \bar{V}/\hbar\omega_D \tag{5.2}$$

Here ω^* is the characteristic frequency of the dissipative system (see (3.29), (3.30) or $\omega^* = \tau_{corr}^{-1}$ (see (3.58)). In the case of the phonon bath described by the Debye model

$$\omega^* = \omega_D \tag{5.3}$$

Without detailed analysis it is clear that the perturbative approach (when terms up to $\bar{V}^2$ are held in the equations of motion) can hardly be justified at very low temperatures and for zero energy gap case ($J_1 - J_2 = 0$). In these cases the first two parameters (5.2) may be infinitely large

$$\xi_1 = \infty\,,\quad \xi_2 \to \infty \tag{5.4}$$

One of the possible solutions of the problem is to derive and to analyze the equation of motion for the electronic subsystem, without making assumptions about the smallness of $\bar{V}$. For this purpose we shall rewrite the Hamiltonian (4.8) (without term G) as follows

$$\mathcal{H} = J_1 n_1 + J_2 n_2 + V \Pi_d (r_+ + r_-) + \sum_k{}' \hbar\omega_k (a_k^+ a_k + \tfrac{1}{2}) + r_+ V(\Pi^+ - \Pi_d) + r_- V(\Pi - \Pi_d) = E + F + W \tag{5.5}$$

where Π_d is the diagonal part of the operator Π (4.3) in the representation in which the operators

$$n_k = a_k^+ a_k$$

are diagonal. It can be easily shown that this operator is equal to

$$\Pi_d = \Pi_d^* = \exp\left[-\frac{1}{2}\sum_k{}' |b_k|^2 (2n_k + 1)\right] \tag{5.6}$$

Now the Hamiltonian of the dynamic (electronic) subsystem has the form (instead of (4.9))

$$E = n_1 J_1 + n_2 J_2 + V_d (r_+ + r_-) = \frac{1}{2}(J_1 + J_2) + r_3 (J_2 - J_1) + 2V_d r_1 \tag{5.7}$$

where

$$V_d = V\Pi_d = V \exp\left[-\frac{1}{2}\sum_k{}' b_k^2 (2n_k + 1)\right] \tag{5.8}$$

The Hamiltonian of the dissipative system is the same as in the preceding Chapter IV (see (4.10), while the interaction energy obtains the form

$$W = r_+ V(\Pi^+ - \Pi_d) + r_- V(\Pi - \Pi_d) \tag{5.9}$$

Further on, we will assume this quantity to be a small perturbation energy. At the same time we will not impose any restrictions on the V_d in (5.7). As we will see later, this enables us to consider the cases when the first two parameters (5.2) ξ_1, and ξ_2 may have arbitrary meanings. Exact meaning of the approximations connected with smallness of W will be analyzed in the next Section.

Now, we apply the general theory of Chapter III (see (3.52)) to the Hamiltonian (5.5). We shall assume, for simplicity's sake that terms K^{VG} in (3.52) may be omitted. Then, according to (3.53), (3.54), (3.45) and (3.67) we have

$$\frac{d\sigma_{mm}}{dt} = -\frac{i}{\hbar}\left[\bar{E},\sigma\right]_{mn} - \sum_{m'n'}{}' \int_0^t K_{mn}^{m'n'}(\tau)\,\sigma_{m'n'}(t-\tau)\,d\tau \tag{5.10}$$

where

$$K_{mn}^{m'n'}(\tau) = \sum_{\alpha\alpha'}{}' K_{m\alpha n\alpha}^{m'\alpha' n'\alpha'}(\tau)\, P_\alpha \quad , \quad \bar{E} = \sum_{\alpha}{}' P_\alpha E_\alpha$$

$$P_\alpha = \exp\left[-F_\alpha/k_BT\right] / Tr\left[\exp\left(-F'/k_BT\right)\right] \tag{5.11}$$

is the equilibrium density matrix of the phonon subsystem. We recall that Latin indices refer to the dynamic subsystem, and Greek indices refer to the dissipative (phonon) subsystem. The Liouville type operator K is equal to

$$K = DL_1 \exp\left[-i\tau(1-D)L_0\right](1-D)L_1 \tag{5.12}$$

and

$$L_1\rho = i\hbar^{-1}[W,\rho]\,,\quad L_0\rho = i\hbar^{-1}[H_0,\rho] =$$
$$= \hbar^{-1}[E,\rho] + \hbar^{-1}[F,\rho] \equiv L^E + L^F \tag{5.13}$$

Now, it is clear that the operator $\exp[-i\tau(1-D)L_o]$ may be presented in the form

$$\exp\left[-i\tau(1-D)L_o\right] = R(\tau)\exp\left[-i\tau(1-D)L^F\right] \tag{5.14}$$

Since, the Hamiltonian E (5.7) coincides with that of (1.44) (after substituting $E_1 \to J_1$; $E_2 \to J_2$) the operator $R(\tau)$ coincides with that of (1.45), (1.51). From (1.51), (5.11), (5.12) the kinetic coefficients, $K_{mn}^{m'n'}$, characterizing the relaxation in (5.10) may be obtained

$$K_{11}^{11}(t) = \bar{V}^2 \hbar^{-2} \Big[\big(e^{F(t)} + e^{F^*(t)} - 2\big)\big(\cos^2(\tfrac{1}{2}\Omega t) - b_3^2 \sin^2(\tfrac{1}{2}\Omega t)\big) +$$

$$+ 2i\cos(\tfrac{1}{2}\Omega t)\sin(\tfrac{1}{2}\Omega t)\, b_3 \big(e^{F} - e^{F^*}\big) - \sin^2(\tfrac{1}{2}\Omega t)\, b_1^2 \big(e^{-F} + e^{-F^*} - 2\big)\Big] \simeq$$

$$\simeq \bar{V}^2 \hbar^{-2} \Big[(F + F^*)\big(\cos^2(\tfrac{1}{2}\Omega t) - (b_3^2 - b_1^2)\sin^2(\tfrac{1}{2}\Omega t)\big) + \tag{5.15}$$

$$+ 2i\cos(\tfrac{1}{2}\Omega t)\sin(\tfrac{1}{2}\Omega t)\, b_3\, (F - F^*)\Big]$$

$$K_{11}^{22}(t) = \bar{V}^2 \hbar^{-2} \Big[-\big(\cos^2(\tfrac{1}{2}\Omega t) - b_3^2 \sin^2(\tfrac{1}{2}\Omega t)\big)\big(e^{F} + e^{F^*} - 2\big) -$$

$$- 2i\, b_3 \cos(\tfrac{1}{2}\Omega t)\sin(\tfrac{1}{2}\Omega t)\big(e^{F^*} - e^{F}\big) + \sin^2(\tfrac{1}{2}\Omega t)\, b_1^2 \big(e^{-F} + e^{-F^*} - 2\big)\Big]$$

$$\simeq \bar{V}^2 \hbar^{-2} \Big[-\big(\cos^2(\tfrac{1}{2}\Omega t) - (b_3^2 - b_1^2)\sin^2(\tfrac{1}{2}\Omega t)\big)(F + F^*) + \tag{5.16}$$

$$+ 2i\, b_3 \cos(\tfrac{1}{2}\Omega t)\sin(\tfrac{1}{2}\Omega t)\,(F - F^*)\Big]$$

$$K_{11}^{12}(t) = \bar{V}^2 \hbar^{-2} \Big[-\big(i\cos(\tfrac{1}{2}\Omega t)\sin(\tfrac{1}{2}\Omega t)\, b_1 - \sin^2(\tfrac{1}{2}\Omega t)\, b_3 b_1\big) \times$$

$$\times \big(e^{-F} + e^{-F^*} - 2\big) - \big(i\cos(\tfrac{1}{2}\Omega t)\, b_1 + \sin^2(\tfrac{1}{2}\Omega t)\, b_3 b_1\big)\big(e^{F^*} + e^{F} - 2\big)\Big]$$

$$\simeq -\bar{V}^2 \hbar^{-2}\, 2\sin^2(\tfrac{1}{2}\Omega t)\, b_3 b_1\, (F + F^*) \tag{5.17}$$

$$K_{11}^{21}(t) = \bar{V}^2 \hbar^{-2} \Big[\Big(i\cos\big(\tfrac{1}{2}\Omega t\big)\sin\big(\tfrac{1}{2}\Omega t\big)\, b_1 - \sin^2\big(\tfrac{1}{2}\Omega t\big) b_3 b_1 \Big) \times$$

$$\times \left(e^{F} + e^{F^*} - 2 \right) + \Big(i\cos\big(\tfrac{1}{2}\Omega t\big)\sin\big(\tfrac{1}{2}\Omega t\big)\, b_1 + \sin^2\big(\tfrac{1}{2}\Omega t\big) b_3 b_1 \Big) \tag{5.18}$$

$$\left(e^{-F} + e^{-F^*} - 2 \right) \Big] \simeq$$

$$\simeq - \bar{V}^2 \hbar^{-2}\, 2\sin^2\big(\tfrac{1}{2}\Omega t\big)\, b_3 b_1 \left(F + F^* \right)$$

$$K_{12}^{11}(t) = \bar{V}^2 \hbar^{-2} \Big[-2 \Big(i\cos\big(\tfrac{1}{2}\Omega t\big)\sin\big(\tfrac{1}{2}\Omega t\big)\, b_1 - \sin^2\big(\tfrac{1}{2}\Omega t\big) b_1 b_3 \Big) \left(e^{F} - 1 \right)$$

$$- 2 \Big(i\cos\big(\tfrac{1}{2}\Omega t\big)\sin\big(\tfrac{1}{2}\Omega t\big) + \sin^2\big(\tfrac{1}{2}\Omega t\big) b_1 b_3 \Big) \left(e^{F^*} - 1 \right) \Big] \simeq \tag{5.19}$$

$$\simeq \bar{V}^2 \hbar^{-2} \Big[2i\cos\big(\tfrac{1}{2}\Omega t\big)\sin\big(\tfrac{1}{2}\Omega t\big)\, b_1 \left(F - F^* \right) - 2\sin^2\big(\tfrac{1}{2}\Omega t\big) b_1 b_3 \left(F + F^* \right) \Big]$$

$$K_{12}^{22}(t) = \bar{V}^2 \hbar^{-2} \Big[2 \Big(i\cos\big(\tfrac{1}{2}\Omega t\big)\sin\big(\tfrac{1}{2}\Omega t\big)\, b_1 + \sin^2\big(\tfrac{1}{2}\Omega t\big) b_1 b_3 \Big) \left(e^{F} - 1 \right)$$

$$+ 2 \Big(i\cos\big(\tfrac{1}{2}\Omega t\big)\sin\big(\tfrac{1}{2}\Omega t\big)\, b_1 - \sin^2\big(\tfrac{1}{2}\Omega t\big) b_1 b_3 \Big) \left(e^{-F^*} - 1 \right) \Big] \tag{5.20}$$

$$\simeq \bar{V}^2 \hbar^{-2} \Big[2i\cos\big(\tfrac{1}{2}\Omega t\big)\sin\big(\tfrac{1}{2}\Omega t\big)\, b_1 \left(F - F^* \right) + 2 b_1 b_3 \sin^2\big(\tfrac{1}{2}\Omega t\big) \left(F + F^* \right) \Big]$$

$$K_{12}^{12}(t) = \bar{V}^2 \hbar^{-2} \Big[\Big(\cos^2\big(\tfrac{1}{2}\Omega t\big) + \sin^2\big(\tfrac{1}{2}\Omega t\big) \left(b_3^2 - b_1^2 \right) \Big) \left(e^{F} + e^{F^*} - 2 \right) \Big]$$

$$\simeq \bar{V}^2 \hbar^{-2} \Big[\Big(\cos^2\big(\tfrac{1}{2}\Omega t\big) + \sin^2\big(\tfrac{1}{2}\Omega t\big) \left(b_3^2 - b_1^2 \right) \Big) \left(F + F^* \right) \Big] \tag{5.21}$$

$$K_{12}^{21}(t) = \bar{V}^2\hbar^{-2}\left[-\left(\cos^2(\tfrac{1}{2}\Omega t) + \sin^2(\tfrac{1}{2}\Omega t)\,(b_3^2 - b_1^2)\right)\left(e^{-F} + e^{F^*} - 2\right)\right.$$

$$\simeq \bar{V}^2\hbar^{-2}\left[\left(\cos^2(\tfrac{1}{2}\Omega t) + \sin^2(\tfrac{1}{2}\Omega t)\,(b_3^2 - b_1^2)\right)(F + F^*)\right] \qquad (5.22)$$

Here b_1, b_3, are defined by relations (1.49), (1.50)*)

$$b_1 = 2\bar{V}\left[(E_1 - E_2)^2 + 4\bar{V}^2\right]^{-1/2}, \quad b_3 = (E_2 - E_1)\left[(E_1 - E_2)^2 + 4\bar{V}^2\right]^{-1/2}$$

$$b_1^2 + b_3^2 = 1 \qquad (5.23)$$

$$\Omega = \hbar^{-1}\left[(E_1 - E_2)^2 + 4\bar{V}^2\right]^{1/2}$$

F(t) is defined by (4.24),

$$F(t) = \sum_k |\eta_k|^2\left[\bar{n}_k e^{-i\omega_k t} + (\bar{n}_k + 1)\, e^{i\omega_k t}\right] = \int_{-a}^{a} \rho(\omega) e^{-i\omega t}\, d\omega \qquad (5.24)$$

$$\bar{V} = V \exp\left[-\tfrac{1}{2}F(0)\right] = V \exp\left[-\sum_k |\eta_k|^2(\bar{n}_k + \tfrac{1}{2})\right] \qquad (5.25)$$

and $\rho(\omega)$ is defined by (4.25). The sign " " of approximate equation in the expressions (5.15)-(5.22) is used in the case when F(t) may be considered as a small quantity. This is valid in the case of small electron-phonon coupling and/or, as we will see later, also for arbitrary electron-phonon coupling but for low temperatures.

*) Rigorously speaking, the quantity $\bar{V}$ appearing in Ω in (5.15)-(5.22), as a result of the averaging procedure (5.11), (5.12), does not coincide exactly with (5.25). The exact coincidence would take place at T = 0. It means that in low temperature regions, approximate coincidence would take place. At high temperatures, as it will be seen from the below, exact meanings of Ω in the formulas (5.15)-(5.22) are not essential.

The equation (5.10) and expressions (5.15)-(5.23) solve, in principle, the problem of time evolution of dynamic subsystems with arbitrary interaction energy V. The equations of motion have the form

$$\dot{\sigma}_{11} = -i\bar{V}\hbar^{-1}(\sigma_{21}-\sigma_{12}) - \int_0^t \Big[K_{11}^{12}(\tau)\sigma_{12}(t-\tau) + K_{11}^{21}(\tau)\sigma_{21}(t-\tau)$$
$$+ K_{11}^{11}(\tau)\sigma_{11}(t-\tau) + K_{11}^{22}(\tau)\sigma_{22}(t-\tau)\Big] d\tau = -\dot{\sigma}_{22} \tag{5.26}$$

$$\dot{\sigma}_{12} = i\bar{V}\hbar^{-1}(\sigma_{11}-\sigma_{22}) - i\omega_{12}\sigma_{12} - \int_0^t \Big[K_{12}^{11}(\tau)\sigma_{11}(t-\tau) +$$
$$+ K_{12}^{22}(\tau)\sigma_{22}(t-\tau) + K_{12}^{12}(\tau)\sigma_{12}(t-\tau) + K_{12}^{21}(\tau)\sigma_{21}(t-\tau)\Big] d\tau = \dot{\sigma}_{21}^{*} \tag{5.27}$$

Further on it will also be convenient to use the equations for the variables (4.120)

$$r = \sigma_{11} - \sigma_{22}\ ,\quad \rho = \sigma_{21} - \sigma_{12}\ ,\quad \sigma = \sigma_{12} + \sigma_{21}\ ,\quad \sigma_{11} + \sigma_{22} = 1$$

These equations have the form*)

$$\dot{r} = -2i\bar{V}\hbar^{-1}\rho - \int_0^t \Big[K_{11}^{12}(\tau) + K_{11}^{21}(\tau)\Big]\sigma(t-\tau)\,d\tau - \int_0^t \Big[K_{11}^{21}(\tau) - K_{11}^{12}(\tau)\Big]\times$$
$$\times \rho(t-\tau)\,d\tau - \int_0^t \Big[\Big(K_{11}^{11}(\tau) - K_{11}^{22}(\tau)\Big) r(t-\tau) + \Big(K_{11}^{11}(\tau) + K_{11}^{22}(\tau)\Big)\Big] d\tau \tag{5.28}$$

$$\dot{\rho} = -2i\bar{V}\hbar^{-1} r + i\omega_{12}\sigma \tag{5.29}$$

$$\dot{\sigma} = i\omega_{12}\rho - \int_0^t \Big[2K_{12}^{12}(\tau)\sigma(t-\tau) + \Big(K_{12}^{11}(\tau) + K_{12}^{22}(\tau)\Big)\Big] d\tau$$
$$- \int_0^t \Big[K_{12}^{11}(\tau) - K_{12}^{22}(\tau)\Big] r(t-\tau)\,d\tau \tag{5.30}$$

Before proceeding further we shall verify that the solution of these equations have the correct equilibrium result as their infinite time limit.

---

*) Approximate relations (5.15)-(5.22) have been taken into account. According to them $K_{12}^{12} = K_{12}^{21}$ and $K_{12}^{mn*} = K_{12}^{mn}$.

The site representation which we use does not diagonalize the unperturbed Hamiltonian E (5.7). In the representation in which E is diagonal, the density matrix $\tilde{\sigma}$ at $t \to \infty$ should be diagonal and satisfy Boltzmann relationship:

$$\tilde{\sigma}_{11} = \left[1 + \exp\left(-\hbar\Omega/k_B T\right)\right]^{-1}$$
$$\tilde{\sigma}_{22} = \exp\left(-\hbar\Omega/k_B T\right)\left[1 + \exp\left(-\hbar\Omega/k_B T\right)\right]^{-1} \tag{5.31}$$

Performing transformation to the site representation, one can get, in the site representation, the equilibrium solution:

$$r(\infty) = b_3\left[2n(\Omega) + 1\right]^{-1} = b_3 \coth \frac{\hbar\Omega}{2k_B T}$$
$$\sigma(\infty) = b_1\left[2n(\Omega) + 1\right]^{-1} = b_1 \coth \frac{\hbar\Omega}{2k_B T} \tag{5.32}$$
$$\rho(\infty) = 0$$

(where Ω is defined by (5.23)). To find the equilibrium solutions of the equations (5.28)-(5.30) we will make Laplace transformation and use the theorem determining the asymptotic behavior of some function f(t)

$$f(\infty) = \lim_{p \to 0} p\hat{f}(p) \tag{5.33}$$

where $\hat{f}(p)$ is the Laplace transform of the function f(t).

The Laplace transform of the equations (5.28)-(5.30) have the form

$$p\hat{r}(p) = r(0) - 2i\bar{V}\hbar^{-1}\hat{\rho}(p) + \left[\hat{K}^{12}_{11}(p) + \hat{K}^{21}_{11}(p)\right]\hat{\sigma}(p)$$
$$- \left[\hat{K}^{21}_{11}(p) - \hat{K}^{12}_{11}(p)\right]\hat{\rho}(p) - \left[\hat{K}^{11}_{11}(p) - \hat{K}^{22}_{11}(p)\right]\hat{r}(p)$$
$$- \left[\hat{K}^{11}_{11}(p) + K^{22}_{11}(p)\right]p^{-1} \tag{5.34}$$

$$p\hat{\rho}(p) = \rho(0) - 2i\bar{V}\hbar^{-1}\hat{\rho}(p) + i\omega_{12}\hat{\sigma}(p) \tag{5.35}$$

$$\begin{aligned} p\hat{\sigma}(p) &= \sigma(0) + i\omega_{12}\hat{\rho}(p) - 2\hat{K}^{12}_{12}(p)\hat{\sigma}(p) \\ &- \left[\hat{K}^{11}_{12}(p) + K^{22}_{12}(p)\right]p^{-1} - \left[\hat{K}^{11}_{12}(p) - \hat{K}^{22}_{12}(p)\right]\hat{r}(p) \end{aligned} \tag{5.36}$$

The coefficients $K^{m'n'}_{mn}(0)$ determine the asymptotic behavior of r,σ,ρ. Using the approximation in which $F(t)$ is small in the expressions (5.15)-(5.22), we obtain

$$\hat{K}^{11}_{12}(0) + \hat{K}^{22}_{12}(0) = 2\pi\bar{V}^2\hbar^{-2}b_1\left[\rho(\Omega) - \rho(-\Omega)\right] \tag{5.37}$$

$$K^{11}_{12}(0) - \hat{K}^{22}_{12}(0) = 2\pi\bar{V}^2\hbar^{-2}b_1 b_3\left[\rho(\Omega) + \rho(-\Omega) - 2\rho(0)\right] \tag{5.38}$$

$$\hat{K}^{11}_{11}(0) + \hat{K}^{22}_{11}(0) = 2\pi\bar{V}^2\hbar^{-2}b_3\left[\rho(\Omega) - \rho(-\Omega)\right] \tag{5.39}$$

$$\hat{K}^{11}_{11}(0) - \hat{K}^{22}_{11}(0) = 2\pi\bar{V}^2\hbar^{-2}\left[b_3^2\left(\rho(\Omega) + \rho(-\Omega)\right) + 2b_1^2\rho(0)\right] \tag{5.40}$$

$$2\hat{K}^{12}_{12}(0) = 2\pi\bar{V}^2\hbar^{-2}\left[b_1^2\left(\rho(\Omega) + \rho(-\Omega)\right) + 2b_3^2\rho(0)\right] \tag{5.41}$$

$$\hat{K}^{12}_{11}(0) + \hat{K}^{21}_{11}(0) = 2\pi\bar{V}^2\hbar^{-2}b_3 b_1\left[\rho(\Omega) + \rho(-\Omega) - 2\rho(0)\right] \tag{5.42}$$

where $\rho(\omega)$ is defined by (4.25). Solving the equations (5.34)-(5.42) and using the theorem (5.33), one can verify that $r(t)$, $\rho(t)$, $\sigma(t)$ have correct asymptotic meanings (5.32).

The equations (5.26), (5.27), in various specific cases, may be essentially simplified. They may be reduced to master equations, provided the condition (see also (4.17))

$$|J_1 - J_2| \gg \bar{V}, \hbar T_{dyn}^{-1} \tag{5.43}$$

is satisfied. (Here T_{dyn} is the characteristic time of relaxation of the dynamic subsystem). In this case

$$\Omega \simeq \omega_{12}, \quad b_1 \simeq 0, \quad b_3 \simeq 1$$

and the equation (5.26) may be reduced to

$$\dot{\sigma}_{11} = - \int_0^t [K_{11}^{11}(\tau)\sigma_{11}(t-\tau) + K_{11}^{22}(\tau)\sigma_{22}(t-\tau)]\, d\tau = -\dot{\sigma}_{22} \tag{5.44}$$

In the Markovian limit (see (3.58)), assuming $K_{11}^{11}(\tau)$, $K_{11}^{22}(\tau)$ are decaying much faster than σ_{11}, σ_{22}, we come to the master equations

$$\dot{\sigma}_{11} = -(W_{12}\sigma_{11} - W_{21}\sigma_{22}) = -\dot{\sigma}_{22} \tag{5.45}$$

where

$$W_{12} = \int_0^\infty K_{11}^{11}(\tau)\, d\tau = \bar{V}^2 \hbar^{-2} \int_{-\infty}^{\infty} e^{-i\omega_{12}\tau} \exp(F(\tau))\, d\tau \tag{5.46}$$

and

$$W_{21} = \bar{V}^2 \hbar^{-2} \int_{-\infty}^{\infty} e^{i\omega_{12}\tau} \exp(F(\tau))\, d\tau \tag{5.47}$$

and $F(\tau)$ is defined by (5.24). Thus, in this case, our general equations, valid for arbitrary V, are reduced to the equations (4.18), (4.19), (4.20) obtained in perturbative approach, V being considered as a small quantity.

We will examine, now, the opposite case

$$\bar{V} \gg |J_1 - J_2| \tag{5.48}$$

In particular, we will consider the symmetric case of the zero energy gap

$$\omega_{12} = 0 \tag{5.49}$$

when parameter (5.2)

$$\xi_1 = \bar{V} / |J_1 - J_2| = \infty \tag{5.50}$$

In the case (5.49) the equations (5.28)-(5.30), (5.15), (5.22) obtain the form

$$\dot{r} = -i\Omega_o \rho - \int_0^t \left[K_{11}^{11}(\tau) - K_{11}^{22}(\tau) \right] r(t-\tau)\, d\tau \tag{5.51}$$

$$\dot{\rho} = -i\Omega_o r \tag{5.52}$$

$$\dot{\sigma} = -\int_0^t 2K_{12}^{12}(\tau)\,\sigma(t-\tau)\, d\tau - \int_0^t \left[K_{12}^{11}(\tau) + K_{12}^{22}(\tau) \right] d\tau \tag{5.53}$$

where

$$K_{12}^{11}(\tau) + K_{12}^{22}(\tau) = \bar{V}^2 \hbar^{-2} \left(e^{i\Omega_o\tau} - e^{-i\Omega_o\tau} \right) \left(F(\tau) - F^*(\tau) \right) \tag{5.54}$$

$$2K_{12}^{12}(\tau) = \bar{V}^2 \hbar^{-2} \left(e^{i\Omega_o\tau} + e^{-i\Omega_o\tau} \right) \left(F(\tau) + F^*(\tau) \right) \tag{5.55}$$

$$K_{11}^{11}(\tau) - K_{11}^{22}(\tau) = 2\bar{V}^2 \hbar^{-2} \left(F(\tau) + F^*(\tau) \right) \tag{5.56}$$

and

$$\Omega_o = 2\bar{V}/\hbar \tag{5.57}$$

Here we have used the approximate relations (5.15)-(5.22) when F(t) may be considered as a small quantity in the exponents. Performing an inspection of the expansion of the exponents of $e^F(t)$ type, similar to that in (4.109), it is easy to verify that use of the approximate relations (5.15)-(5.22) (with $\omega_{12} = 0$) is justified under the following conditions

$$\left(\bar{V}/\hbar\omega_D \right)^2 \left(E_r/\hbar\omega_D \right) \ll 1 \quad \text{for } k_B T \lesssim \bar{V} \tag{5.58}$$

and at higher temperatures one can use relations (4.115)-(4.117).

Let us first examine the equation (5.53). In the Markovian approximation we will make an ansatz that the variation of σ is much slower than that of K(τ). Then, the equation for σ would take the form

$$\dot{\sigma} = -\gamma(\sigma - \sigma_o) \tag{5.59}$$

Here

$$\gamma = \int_0^\infty 2 K_{12}^{12}(\tau)\, d\tau = 2\pi \bar{V}^2 \hbar^{-2} f(\Omega_0)\, \eta^2(\Omega_0) \left[2n(\Omega_0) + 1 \right] \tag{5.60}$$

$$\sigma_0 = \left[2n(\Omega_0) + 1 \right]^{-1} \tag{5.61}$$

where $f(\omega)$ is the frequency distribution, $\eta^2(\omega)$ and $n(\omega)$ have been defined above: (4.25)-(4.27).

To justify our ansatz, we utilize the procedure similar to (3.26)-(3.27) and present first term in the r.h.s. (5.53) in the form

$$-\frac{1}{2\pi i} \lim_{\varepsilon \to +0} \int_{\varepsilon - i\infty}^{\varepsilon + i\infty} \hat{K}_{12}^{12}(p)\, \hat{\sigma}(p)\, e^{pt}\, dp \tag{5.62}$$

where "^" is the sign of Laplace transform. The expression (5.62) may be approximately cast in the form

$$-\hat{K}_{12}^{12}(0) \frac{1}{2\pi i} \int_{\varepsilon - i\infty}^{\varepsilon + i\infty} \hat{\sigma}(p)\, e^{pt}\, dp = -\hat{K}_{12}^{12}(0)\, \sigma(t) \tag{5.63}$$

provided the variation of $\hat{K}_{12}^{12}(p)$ near $p \simeq 0$ is slow in comparison with that of $\hat{\sigma}(p)$. According to (5.59) the interval of essential variation of p in the integral (5.63) is of the order of magnitude

$$|\Delta p| \sim \gamma$$

To estimate the interval of variation $\hat{K}(p)$ we should assume some model of the frequency distribution $f(\omega)$. It is obvious that at low temperatures the main contribution comes from low frequency acoustical phonons. According to Chapter 4.2, in a quite general case, the low frequency behavior of $f(\omega)$ and $\eta^2(\omega)$, may be described by the formulae (4.54) and (4.56) respectively. Using these formulae, (5.55) and (4.25) it is easy to check that $\hat{K}_{12}^{12}(p)$ essentially varies in the following intervals (near $p \simeq 0$)

$$|\Delta p| \sim \tau_c^{-1} \simeq \begin{cases} \Omega_0 & \text{for } k_B T \lesssim \hbar\Omega_0 < \hbar\omega_D \\ k_B T/\hbar & \text{for } \hbar\omega_D > k_B T > \hbar\Omega_0 \\ \omega_D & \text{for } k_B T > \hbar\omega_D,\ \hbar\Omega_0 \end{cases} \tag{5.64}$$

Thus the condition of transition to the equation (5.59) (compare with condition (3.58)!) has the form

$$\gamma\tau_c \simeq \begin{cases} (\bar V/\hbar\omega_D)^2 (E_r/\hbar\omega_D) \ll 1 & \text{for } \begin{cases} k_B T < \bar V \\ \hbar\omega_D > k_B T > \bar V \end{cases} \\ (\bar V/\hbar\omega_D)^2 (E_r/\hbar\omega_D)(k_B T/\hbar\omega_D) \ll 1 & \text{for } k_B T > \bar V, \hbar\omega_D \end{cases} \tag{5.65}$$

To analyze the equations (5.51), (5.52) we introduce new variables

$$R_\pm = r \mp \rho = R^{\pm} e^{i\Omega_0 t} \tag{5.66}$$

In these variables the equations (5.51), (5.52) take the form

$$\begin{aligned} \dot R_0^{\pm} = &-\int_0^t \left[K_{11}^{11}(\tau) - K_{11}^{22}(\tau)\right] e^{\mp i\Omega_0\tau} R_0^{\pm}(t-\tau)\, d\tau \\ &- \int_0^t \left[K_{11}^{11}(\tau) - K_{11}^{22}(\tau)\right] e^{\pm i\Omega_0\tau} R^{\mp}(t-\tau)\, e^{\mp 2i\Omega_0 t} \end{aligned} \tag{5.67}$$

Using the same arguments which have led us to the equation (5.59), and under the same conditions (5.59), we get

$$\dot R_\pm = \pm i(\Omega_0 + \delta) R_\pm - \gamma R_\pm \tag{5.68}$$

where

$$\delta = 2\bar V^2 \hbar^{-2} \lim_{\varepsilon \to +0} \int_0^a d\omega\, f(\omega)\, \xi^2(\omega) \left(2n(\omega)+1\right) \times$$

$$\times \left\{(-\Omega_0 - \omega)\left[(\Omega_0+\omega)^2 + \varepsilon^2\right]^{-1} + (-\Omega_0+\omega)\left[(\Omega_0-\omega)^2+\varepsilon^2\right]^{-1}\right\}$$

(It is easy to show that the oscillating terms in (5.67) may be thrown out, if $\gamma \ll \Omega_0$).

Results of this section may be summarized as follows. We have started from the Hamiltonian (4.8), widely used for the description of various radiationless processes. The approximation, which is usually applied, considers the interaction energy V in this Hamiltonian as a small parameter. The arguments were given, that at low temperatures and for small energy gaps $J_1 - J_2$, such approximation may not be satisfactory. To overcome the limitation of small V, the Hamiltonian (4.8) has been presented in the form (5.5) with perturbation energy W (5.9). The equations of motion for the density matrix of the dynamic subsystem σ have been derived (5.10). These equations are valid for arbitrary values of $\bar{V}$, but for small W only. It has been shown that the equations (5.10) have correct asymptotic behavior ($t \to \infty$), and that for large energy gaps (5.43) these equations may be reduced to usual master equations.

Then the case of zero energy gap (5.49) has been analyzed. The equations which have been derived - (5.59) and (5.68), at large temperatures

$$k_B T \gg \bar{V} \tag{5.69}$$

coincide with those derived earlier (4.121, (4.122). It is easy to check that from (4.14) and (5.60), it follows that

$$\gamma \simeq 2w \tag{5.70}$$

provided condition (5.69) is fulfilled. In the low temperature region

$$k_B T \lesssim \bar{V} \tag{5.71}$$

both equations and coefficients substantially differ from those derived in the preceding Chapter (4.121), (4.122). The qualitative difference is in the fact that the quantity

$$\sigma = \sigma_{12} + \sigma_{21}$$

characterizing so-called coherence does not vanish when $t \to \infty$. Its asymptotic value equals σ_o (5.11). It is connected with the previously mentioned circumstance that our (site) representation does not diagonalize the unperturbed Hamiltonian (5.7) and $\sigma_o \neq 0$ is necessary to provide correct equilibrium behavior (5.31) of the equation (5.59). At the absolute zero temperature T = 0, the parameter

$$\xi_2 = \bar{V}/k_B T \to \infty$$

However, the decay constant γ (5.60) in the equations (5.59), (5.68) has finite, non-zero value. According to (5.60), (4.54), (4.56) at the absolute zero temperature $T = 0$ ($n(\Omega_o)=0$), the decay constant γ is equal

$$\gamma = 12\pi \left(\bar{V}/\hbar\omega_D\right)^3 \left(E_r/\hbar\right) \tag{5.72}$$

Cubic dependence on $\bar{V}$ has a simple explanation. At T = 0, the only process contributing to relaxation from non-stationary state to equilibrium is the spontaneous emission of acoustical phonons. Intensity of such process is proportional to Ω_o^3, where Ω_o is the frequency of emitted phonons. According to (5.57), Ω_o is proportional to $\bar{V}$ and thus γ should be proportional to $\bar{V}^3$. Such dependence on $\bar{V}$ cannot be obtained by perturbative approach, $\bar{V}$ serving a small parameter.

In conclusion, it is worthwhile mentioning that in the temperature region considered here (5.71), the transition between two states 1 and 2 of the dynamic subsystem, has the character of decaying quantum beats with frequency

$$\Omega_0 \gg \gamma$$

5.2 Validity conditions of perturbative approach

The important question is, what are the conditions enabling us to use the perturbative approach both here - at low temperatures (5.71), and in other temperature regions considered in Chapter IV.

As a matter of fact we have used two different approximations: one of them connected with expansion of the functions e^F in (5.15)-(5.22), and another one connected with smallness of the perturbation energy W. The first kind of approximation is determined by the relation (5.58) at low temperatures and by (4.115)-(4.117) at higher temperatures. The smallness of the perturbation energy W means that the parameter

$$\gamma \tau_c \ll 1 \tag{5.73}$$

where τ_c - characteristic time of correlation in the dissipative system and it is defined by (5.64). The condition (5.73) is similar to that of (3.58). This condition determines the applicability of the perturbative approach and the Markovian approximation at the same time. It essentially coincides with conditions (1.78), (1.79) limitating use of the transition probability per unit time. Thus in the low temperature region

$$k_B T \lesssim \bar{V}$$

according to (5.65), the condition of the applicability of the perturbative approach (smallness of W) obtains the form

$$(\bar{V}/\hbar\omega_D)^2 (E_r/\hbar\omega_D) \ll 1 \tag{5.74}$$

Coming back to three parameters (5.2) introduced in the preceding section 5.1 we see that the theory developed there does not impose any limitations on the first two parameters

$$\xi_1 = |\bar{V}|/|J_1 - J_2| \,, \quad \xi_2 = \bar{V}/k_B T$$

On the contrary, both of them may have arbitrary values including $\xi_1 = \infty$; $\xi_2 = \infty$, for zero-energy gap and absolute zero temperature. Then if the electron-phonon coupling parameter is small

$$E_r/\hbar\omega_D << 1$$

the theory does not impose a substantial restriction on the third parameter $\xi_3=\bar{V}/\hbar\omega_D$ as well. It is important to stress that for the case of the strong electron-phonon coupling

$$E_r/\hbar\omega_D >> 1 \tag{5.75}$$

condition (5.74) would be fulfilled for very broad range of unaveraged ("undressed") interaction energies V. To show this, it is enough to recall that according to (5.1), (4.123) condition (5.74) may be rewritten as

$$(V/\hbar\omega_D)^2 (E_r/\hbar\omega_D) \exp\left(-\frac{3}{2} E_r/\hbar\omega_D\right) << 1 \tag{5.76}$$

Now, let us consider the temperature region

$$k_B T >> \bar{V}, \quad \xi_2 << 1 \tag{5.77}$$

We will show that this case corresponds to the calculating of the kinetic coefficients $K_{mn}^{m'n'}$ with accuracy up to V^2 terms. It is easy to understand that the characteristic time scale of the coefficients $K_{mn}^{m'n'}(t)$ in (5.15)-(5.22) is

$$\tau_c \lesssim \begin{cases} \hbar/k_B T & \text{for } k_B T < \hbar\omega_D \\ \omega_D^{-1} & \text{for } k_B T > \hbar\omega_D \end{cases} \tag{5.78}$$

In the case of zero energy gap $\omega_{12} = 0$*) the quantity Ω in (5.15)-(5.22) is equal to

$$\Omega_0 = 2\bar{V}/\hbar$$

and from (5.78) it follows that in the expressions (5.15)-(5.22) $\Omega_o t$ is limited by the relations

$$\Omega_0 t \lesssim \begin{cases} \bar{V}/k_B T, & k_B T < \hbar\omega_D \\ \bar{V}/\hbar\omega_D, & k_B T > \hbar\omega_D \end{cases} \tag{5.79}$$

---

*) For large energy gaps (5.43) the perturbative approach has been justified above.

Assuming (5.77) and

$$\xi_3 = \bar{V}/\hbar\omega_D << 1 \tag{5.80}$$

one may neglect $\Omega_o t$ in the exponents (5.15), (5.22). (This conclusion may be verified also by direct calculation of the integrals in $K_{mn}^{m'n'}$ (3.67) ,and expressions of (4.135) type where instead of ω_{12} one should put Ω_o). These considerations justify the perturbative approach used in the preceding chapter, when terms up to V^2 are taken into account, calculating the kinetic coefficients $K_{mn}^{m'n'}$.

In the temperature region (4.125)

$$(E_r/\hbar\omega_D)(k_B T/\hbar\omega_D)^2 >> 1 \tag{5.81}$$

when the saddle point method may be applied, it is easy to get the condition of the perturbative approach in the form (see (5.73), (5.78), (4.136))

$$\begin{gathered} w\tau_c \sim w\hbar/k_B T \sim (\bar{V}/k_B T)^2 \zeta^{-1/2} e^{\zeta} << 1 \\ \zeta = \int_{-a}^{a} \rho(\omega) \exp\left(\tfrac{1}{2}\hbar\omega/k_B T\right) d\omega >> 1 \end{gathered} \tag{5.82}$$

at $k_B T << h\omega_D$.

This means that parameter $\xi_2 = \bar{V}/k_B T$ should be small, this condition being consistent with our initial assumption.

In the case

$$k_B T >> \hbar\omega_D \tag{5.83}$$

according to (5.78), (4.136), the condition of smallness of $\bar{V}$ takes the form

$$w\tau_c \sim w/\omega_D \sim (\bar{V}/\hbar\omega_D)^2 \zeta^{-1/2} e^{\zeta} << 1 \tag{5.84}$$

which means that parameter

$$\xi_3 = \bar{V}/\hbar\omega_D$$

should be small since $\zeta >> 1$. On the other hand the condition (5.84) may be written as (see (4.139))

$$|V|^2\left[k_BTE_r(\hbar\omega_D)^2\right]^{-1/2}\exp\left(-E_r/4k_BT\right) < 1 \tag{5.85}$$

It is clear that this condition is a generalization of the condition (2.42) of the Landau-Zener approximation. The difference is in the fact that (2.42) corresponds to one degree of freedom, while (5.85) takes into account a continuum of degrees of freedom.

Up till now we have used the Debye model for our estimates. It is obvious that at very low temperatures, the low frequency acoustical phonons should give the main contribution to the rate processes. As it has been mentioned above the low frequency acoustical phonons may be quite satisfactorily modeled by the Debye distribution.

However, at higher temperatures both optical and acoustical phonons may contribute to the rate processes. The optical phonons may be presented by the Einstein model (4.62), (4.63). At high enough temperatures we may again use the expression (4.136). In the Einstein model this expression obtains the form

$$\tfrac{1}{2}\gamma = w = (\bar{V}^2/\hbar^2\omega_0)(2\pi/\zeta)^{1/2}\exp(\zeta)$$

$$\zeta = 2(E_r/\hbar\omega_0)\, n(\omega_0)\exp(\tfrac{1}{2}\hbar\omega_0/k_BT)$$

For $k_BT > h\delta$ (see (4.62)) role of τ_c in (5.73) plays δ^{-1}. Thus the condition of the applicability of the perturbative approach, (smallness of V^2) has the form

$$\gamma\tau_c \sim \gamma/\delta \sim (\bar{V}^2/\hbar^2\omega_0\delta)\,\zeta^{-1/2}\exp(\zeta) \ll 1 \tag{5.86}$$

which implies ($\zeta \gg 1$) that

$$\bar{V}^2/\hbar^2\omega_0\delta \ll 1 \tag{5.87}$$

Thus, we have examined the conditions under which the perturbative approach is valid. We have considered the specific case of zero energy gap. Using the derived in preceding chapter expressions for non-zero energy gap it is easy to deduce similar conditions. It is obvious that these conditions would impose less restrictions on $\bar{V}^2$ since corresponding expressions for w have lower meanings.

VI. ADIABATIC RATE PROCESSES

6.1 Tunneling in condensed media

As we remember (Chapter 2.1), the non-radiative transitions occurring at the same electronic state are called adiabatic transitions. It means that in terms of the Born-Oppenheimer approximation (see Chapter 1.4) the transitions are going on at the same electronic energy hypersurface between two potential energy minima corresponding to two (quasi) stable configurations of nuclei.

We consider, here, a quite general model [47] which may be appropriate for the description of proton transfer, nuclear group transfer, electron transfer accompanied by the transfer of nuclear groups,and other adiabatic rate processes in condensed media.

In the Born-Oppenheimer approximation all such processes may be described by the Hamiltonian

$$\mathcal{H} = T + U(Q) + \frac{1}{2}\sum_k{}' (p_k^2 + \omega_k^2 q_k^2) - \sum_k{}' A_k(Q) q_k \qquad (6.1)$$

Here Q is the set of coordinates of singled out nuclear modes describing the nuclear subsystem (molecule) interacting with its surrounding - a condensed medium. T and U are kinetic and potential energies of the nuclear subsystem; the third term in the Hamiltonian (6.1) describes the condensed medium in the harmonic approximation - as a phonon bath, and the last term is the interaction energy between the nuclear subsystem and condensed medium, the only assumption about this interaction is that it is linear in the coordinates q_k, i.e. the excitations of the surrounding condensed medium are small enough.

The potential energy U is supposed to have two minima corresponding to two (quasi) stable configurations of the nuclear subsystem. The condition of the (quasi) stability of the configurations is that they are divided by a barrier which is sufficiently large, and that tunneling through this barrier may be considered as a small perturbation.

Other examples of processes which may be described by the Hamiltonian (6.1) are the following:

Enzyme catalysis plays a central role in all biological systems. As is known [48], enzymatic catalysis goes on through the stage of an enzyme substrate complex, and this stage, as a rule, limits the rate of enzymatic catalysis. At this stage a substrate (or substrates), i.e. a compound undergoing a chemical transformation into product (or products) of reaction, forms together with an enzyme - a protein macromolecule, an intermediate combination - an enzyme-substrate complex. Thus, the Hamiltonian (6.1) may describe the transition: substrate-enzyme complex $\rightleftarrows$ product-enzyme complex. Another example: transition between the degenerate isomer states of optically active molecules embedded in condensed media. Recently, [49,50] this phenomenon and possible experimental implications have been discussed in the literature [47,48].

Transition from one configuration (Fig. 9) (potential well 1) to another (potential well 2) may occur in two ways: (1) tunneling through the potential barrier and (2) overcoming the potential barrier by thermal fluctuations induced by the phonon bath. In this section we would explore the first possibility, i.e. the tunneling.

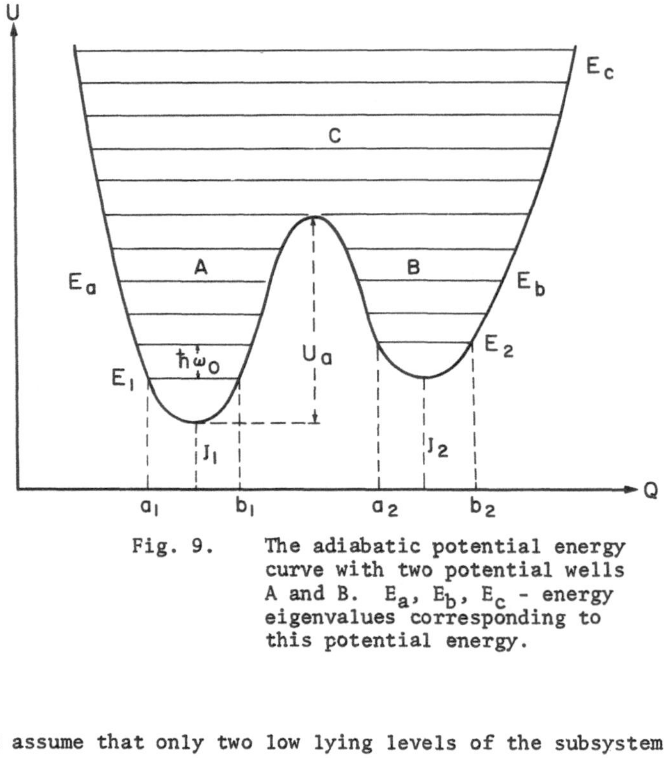

Fig. 9. The adiabatic potential energy curve with two potential wells A and B. E_a, E_b, E_c - energy eigenvalues corresponding to this potential energy.

We would assume that only two low lying levels of the subsystem are essential, i.e. these two levels E_1 and E_2 are well separated from the others and it is possible to neglect excitations to other levels. It may mean, in particular, that the temperature of the bath is small enough

$$k_B T \ll \hbar\omega_0 \tag{6.2}$$

where $h\omega_0$ is the characteristic energy difference between E_1 or E_2 and the first excited state in each well. In this case we are dealing with a two-state system (see Chapter 1.2) and the Hamiltonian of the nuclear subsystem may be represented in the form (1.40)

$$\mathcal{H}_0 = n_1 E_1 + n_2 E_2 + r_+ V_{12} + r_- V_{21} \tag{6.3}$$

where operators n_1, n_2, $r_\pm$ are defined by relations (1.33)-(1.35), (1.41), (1.42) and V_{12}, V_{21} are the matrix element of the effective perturbation energy corresponding to the possibility of tunneling between states 1 and 2. Above (2.14) we have calculated the effective perturbation energy matrix elements for one-dimensional case in the semiclassical approximation

$$V_{12} = V_{21} = V = \frac{1}{2}\hbar\left(\omega_1\omega_2/\pi e\right)^{1/2} \exp\left(-\hbar^{-1}\int_{b_1}^{a_1} |p|\, dQ\right) \tag{6.4}$$

where ω_1, ω_2 are the frequencies characterizing wells 1 and 2; $p = \sqrt{2M(E-U)}$; $E_1 \simeq E_2 = E$. In the case of more than one singled out nuclear modes Q it is reasonable to assume that the matrix elements have the form

$$V_{12} = V_{21}^* = \hbar\omega e^{-\sigma} \tag{6.5}$$

where the effective parameter ω has the order of magnitude ω_0 in (6.2) and the parameter σ is assumed to be much larger than unity

$$\sigma \gg 1 \tag{6.6}$$

Now, taking into account only two levels E_1 and E_2 of the nuclear subsystem the Hamiltonian (6.1) of the whole system may be presented in the form

$$\mathcal{H} = n_1 \mathcal{H}_{11} + n_2 \mathcal{H}_{22} + r_+ \mathcal{H}_{12} + r_- \mathcal{H}_{21} \tag{6.7}$$

where the matrix elements $\mathcal{H}_{ik}$ are taken with the aid of the eigenfunctions $\Psi_1(Q)$, $\Psi_2(Q)$ corresponding to the eigenvalues E_1 and E_2. Thus, the effective Hamiltonian of the whole system (including condensed medium) takes the form

$$\begin{aligned} \mathcal{H} = {} & n_1 E_1 + n_2 E_2 + \frac{1}{2}\sum_k \left(p_k^2 + \omega_k^2 q_k^2\right) - \\ & - n_1 \sum_k \omega_k^2 q_{1k}^0 q_k - n_2 \sum_k \omega_k^2 q_{2k}^0 q_k + \\ & + r_+\left[V_{12} + \sum_k V_{12k} q_k\right] + r_-\left[V_{21} + \sum_k V_{21k} q_k\right] \end{aligned} \tag{6.8}$$

where

$$\omega_k^2 q_{1k}^0 = \int \Psi_1^* A_k(Q) \Psi_1 \, dQ$$

$$\omega_k^2 q_{2k}^0 = \int \Psi_2^* A_k(Q) \Psi_2 \, dQ$$

$$V_{12k} = V_{21k}^* = -\int \Psi_1^* A_k(Q) \Psi_2 \, dQ$$

$$Q = \{Q_1, \ldots, Q_n\}$$

Thus, apart from the terms V_{12k}, V_{21k}, the Hamiltonian (6.8) is isomorphous to that of (2.47) describing the electron transfer. All the analysis of the preceding Chapters IV and V was essentially based on the Hamiltonians (2.47) (4.1). These Hamiltonians coincide with those of (6.8) up to designations.*) Thus, transition between two wells A and B is described by the equations (4.11), (4.12) or (5.10). Analysis of these equations and their coefficients, which was given in the preceding Chapters IV and V is also valid for description of tunneling in condensed media.

At the end of this chapter we would return to specific features of the adiabatic transitions, following from the Hamiltonian (6.8) and the analysis in the preceding chapters.

6.2 Overcoming the potential barrier

In this section we will continue to consider the adiabatic transitions between two potential wells located at the same potential energy hypersurface (see Fig. 9). However, now we will consider the high-temperature transitions [47], when

$$k_B T \gg \hbar\omega_0 \tag{6.9}$$

This is just the opposite case of (6.2). We will assume also the validity of semi-classical approximation for description of the nuclear system. It means that in the limit $\hbar \to 0$ the energy levels of the nuclear system may be classified according to Fig. 9. We have here 3 types of states a,b,c, and within the classical limit, the

*) Omission of the terms V_{12k}, V_{21k} is equivalent to using the Condon approximation in (4.1), i.e. to assume that V does not depend on q.

only allowed transitions are those between a, b, c themselves and between a and c, b and c. In this approximation, a transition from the well A to the well B goes through transitions from states a to c and afterwards from c to b. Thus, we take into account the following transition probabilities $w_{aa'}$, $w_{bb'}$, $w_{cc'}$ and w_{ca}, w_{cb}, w_{ac}, w_{bc} .

The corresponding master equations may be written in the form

$$
\begin{aligned}
\dot{P}_a &= -\sum_{a'}{}' \left(w_{aa'} P_a - w_{a'a} P_{a'}\right) + \sum_{c}{}' \left(w_{ca} P_c - w_{ac} P_a\right) \\
\dot{P}_b &= -\sum_{b'}{}' \left(w_{bb'} P_b - w_{b'b} P_{b'}\right) + \sum_{c}{}' \left(w_{cb} P_b - w_{bc} P_b\right) \\
\dot{P}_c &= \sum_{a}{}' \left(w_{ac} P_a - w_{ca} P_c\right) + \sum_{b}{}' \left(w_{bc} P_b - w_{cb} P_c\right) + \\
&\quad \sum_{c'}{}' \left(w_{cc'} P_c - w_{c'c} P_{c'}\right)
\end{aligned}
\tag{6.10}
$$

where P_a, P_b, P_c are the populations of the states a,b,c respectively.

For the equilibrium surrounding the medium the transition probabilities w_{ab} satisfy the relation (3.70)

$$
w_{ab} = w_{ba} \exp\left[\beta \left(E_a - E_b\right)\right] \tag{6.11}
$$

where E_a is the energy of the microstate a and

$$
\beta = 1/k_B T
$$

We will look for the solution of (6.10) in the form

$$
\begin{aligned}
P_a &= c_A(t)\, e^{-\beta E_a} / Z_A + \eta_a \\
P_b &= c_B(t)\, e^{-\beta E_b} / Z_B + \eta_b \\
P_c &= N_c\, e^{-\beta E_c}
\end{aligned}
\tag{6.12}
$$

where

$$\Sigma'_A = \sum_a{}' e^{-\beta E_a} \quad , \quad \Sigma'_B = \sum_b e^{-\beta E_b} \tag{6.13}$$

are the statistical sums of the states A and B: C_A and C_B are total concentrations of the species A and B respectively

$$C_A = \sum_a{}' P_a \quad , \quad C_B = \sum_b{}' P_b \tag{6.14}$$

The quantities η_a and η_b satisfy the normalization conditions

$$\sum_a{}' \eta_a = \sum_b{}' \eta_b = 0$$

Substituting (6.12) into (6.10) and using (6.11), we get

$$\dot{\eta}_a = \sum_{a'}{}' \left(w_{a'a}\eta_{a'} - w_{aa'}\eta_a \right) - \sum_c{}' w_{ca}\eta_a \exp\left[\beta(E_a - E_c)\right] +$$

$$+ \sum_c{}' w_{ca} N_c e^{-\beta E_c} - \sum_c{}' w_{ca} C_A e^{-\beta E_c}/\Sigma'_A - \dot{C}_A e^{-\beta E_a}/\Sigma'_A \tag{6.15}$$

$$\dot{\eta}_b = \sum_{b'}{}' \left(w_{b'b}\eta_{b'} - w_{bb'}\eta_b \right) - \sum_b{}' w_{cb}\eta_b \exp\left[\beta(E_b - E_c)\right]$$

$$+ \sum_c{}' w_{cb} N_c e^{-\beta E_c} - \sum_c{}' w_{cb} C_B e^{-\beta E_c}/\Sigma'_B - \dot{C}_B e^{-\beta E_b}/\Sigma'_B$$

$$\dot{N}_c = \gamma_{cA} C_A/\Sigma'_A + \gamma_{cB} C_B/\Sigma'_B - (\gamma_{cA} + \gamma_{cB}) N_c -$$

$$- \sum_{c'}{}' w_{cc'}(N_c - N_{c'}) + \sum_a{}' w_{ca}\eta_a \exp\left[\beta(E_a - E_c)\right] +$$

$$+ \sum_b{}' w_{cb}\eta_b \exp\left[\beta(E_b - E_c)\right] \tag{6.16}$$

Here

$$\gamma_{cA} = \sum_a{}' w_{ca} \;, \quad \gamma_{cB} = \sum_b{}' w_{cb} \tag{6.17}$$

Performing summation over a and b in (6.15) we get

$$\dot{c}_A = \sum_c{}' e^{-\beta E_c} \gamma_{cA} \left(N_c - c_A / \Sigma'_A\right) - \sum_{ca}{}' w_{ca} \eta_a \exp\left[-\beta(E_c - E_a)\right]$$

$$\dot{c}_B = \sum_c{}' e^{-\beta E_c} \gamma_{cB} \left(N_c - c_B / \Sigma_B\right) - \sum_{cb}{}' w_{cb} \eta_b \exp\left[-\beta(E_c - E_b)\right] \tag{6.18}$$

Further we assume that

$$k_B T \ll U_a \tag{6.19}$$

where U_a is the activation energy. This means that $\exp(-U_a/kT)$ may be considered as a small parameter. [A similar condition has been assumed above: (2.31)].

It is worthwhile to underline the fact that the condition (6.19) is quite a natural one for the problem of overcoming the potential barrier in the adiabatic case, because only when condition (6.19) is satisfied, do particles spend almost all their time either in well A or in well B. In the opposite case $kT > U_a$ we cannot speak about the particles localized in one of the wells; comparable number of the particles are in the states C.

When the condition (6.19) is satisfied it is easy to verify from (6.15) that η_a and η_b have the order of magnitude

$$\eta_a \sim c_A \exp(-\beta E_c)/\Sigma_A \;, \quad \eta_b \sim c_B \exp(-\beta E_c)/\Sigma'_B \tag{6.20}$$

Using these estimates one can simplify equations (6.16), (6.18), and to omit terms containing η_a, η_b, provided the following conditions are fulfilled*)

$$\sum_a{}' w_{ca} \exp\left[-\beta(E_c - E_a)\right] \Big/ \sum_a{}' w_{ca} \ll 1$$

$$\sum_b{}' w_{cb} \exp\left[-\beta(E_c - E_b)\right] \Big/ \sum_b{}' w_{cb} \ll 1 \tag{6.21}$$

---

*) These conditions may follow from (6.19), provided various w_{ca} are of the same order of magnitude

In this case the equations (6.16), (6.18) obtain the form

$$\dot{N}_c = \gamma_{cA} C_A/\Sigma_A + \gamma_{cB} C_B/\Sigma_B - (\gamma_{cA}+\gamma_{cB}) N_c - \sum_{c'}{}' w_{cc'}(N_c - N_{c'}) \tag{6.22}$$

$$\dot{C}_A = \sum_{c}{}' e^{-\beta E_c} \gamma_{cA} \left(N_c - C_A/\Sigma_A\right) \tag{6.23}$$

$$\dot{C}_B = \sum_{c} e^{-\beta E_c} \gamma_{cB} \left(N_c - C_B/\Sigma_B\right) \tag{6.24}$$

Examining these equations one realizes that the concentrations C_A and C_B are slow variables in comparison with N_c. The derivatives $\dot{C}_A$ and $\dot{C}_B$ contain the extra factor of the order of magnitude $\exp(-\beta U_a)$ which $\dot{N}_c$ does not contain. It means that in the same approximation (6.19) one can use the steady state condition

$$\dot{N}_c = 0$$

Thus we have from (6.22)

$$N_c = (\gamma_{cA}+\gamma_{cB})^{-1}\left(\gamma_{cA} C_A/\Sigma_A + \gamma_{cB} C_B/\Sigma_B\right) + (\gamma_{cA}+\gamma_{cB})^{-1} \sum_{c'} w_{cc'}(N_c - N_{c'}) \tag{6.25}$$

The equations (6.23)-(6.25) solve the problem of time evolution of the total concentrations C_A, C_B in the course of the reaction.

In various limiting cases they may be essentially simplified. The equation (6.25) has the exact solution in the case of symmetrical wells, when

$$\Sigma_A = \Sigma_B = \Sigma \qquad \gamma_{cA} = \gamma_{cB} = \gamma_c$$

and

$$N_c = (C_A + C_B)/2\Sigma$$

The equations (6.23), (6.24), for the concentrations obtain the form

$$\dot{C}_A = -k\,(C_A - C_B) = -\dot{C}_B \tag{6.26}$$

where the reaction rate constant k has the form

$$k = \sum_c \gamma_c e^{-\beta E_c}/2\Sigma = \tfrac{1}{2}\bar{\gamma} \exp[-\beta(F_c - F_A)] \tag{6.27}$$

Here the mean value of the transition probability equals

$$\bar{\gamma} = \sum_c{}' \gamma_c e^{-\beta E_c} \Big/ \sum{}' e^{-\beta E_c} \tag{6.28}$$

and the free energies of the initial state F_A (= F_B) and the transition state F_C are expressed through the statistical sums in a usual way

$$e^{-\beta F_A} = \sum_a e^{-\beta E_a} \quad , \quad e^{-\beta F_C} = \sum_c{}' e^{-\beta E_c} = \sum{}'_C \tag{6.29}$$

The equations (6.23)-(6.25) may also be simplified in two limiting cases.

(a) The relaxation between microstates c of the transiticn state C is much slower than relaxation from the state c to the states a and b

$$\gamma_{cA} , \gamma_{cB} \gg w_{cc'} \tag{6.30}$$

In this case the approximate solution of (6.25) takes the form

$$N_c = (\gamma_{cA} + \gamma_{cB})^{-1} (\gamma_{cA} C_A / \Sigma_A + \gamma_{cB} C_B / \Sigma_B) \tag{6.31}$$

and the equations for the concentrations C_A and C_B obtain the form

$$\dot{C}_A = - k_{AB} C_A + k_{BA} C_B = - \dot{C}_B \tag{6.32}$$

where

$$\begin{aligned} k_{AB} &= \gamma_{ACB} \, e^{-\beta (F_C - F_A)} \\ k_{BA} &= \gamma_{ACB} \, e^{-\beta (F_C - F_B)} \end{aligned} \tag{6.33}$$

Here, again, F_C is the free energy of the transition state, and F_A, F_B are those of the states A and B. The pre-exponential factor in (6.33) is defined by the formula

$$\dot{\gamma}_{ACB} = \left(\sum_c{}' e^{-\beta E_c} \right)^{-1} \sum_c \gamma_{cA} \gamma_{cB} (\gamma_{cA} + \gamma_{cB})^{-1} e^{-\beta E_c} \tag{6.34}$$

It is easy to verify that the ratio of the rate constants (6.33) is equal to the equilibrium constant

$$(C_A / C_B)_{t \to \infty} = K = \frac{k_{BA}}{k_{AB}} = e^{-\beta (F_A - F_B)} \tag{6.35}$$

as it should be.

(b) The relaxation between the microstates c of the transition state C is much faster than relaxation from the transition microstates c to the states a and b

$$\gamma_{cA} , \gamma_{cB} \ll w_{cc'} \tag{6.36}$$

To analyze this case we will express N_c through P_c (see (6.12)) and rewrite (6.25) in the form

$$(\gamma_{cA}+\gamma_{cB})P_c+\sum_{c'}(w_{cc'}P_c-w_{c'c}P_c)=(\gamma_{cA}/\Sigma_A+\gamma_{cB}/\Sigma_B)e^{-\beta E_c} \qquad (6.37)$$

Performing summation over c in this formula we get

$$\sum_c (\gamma_{cA}+\gamma_{cB})P_c=\sum_c(\gamma_{cA}C_A/\Sigma_A+\gamma_{cB}C_B/\Sigma_A)e^{-\beta E_c} \qquad (6.38)$$

One can check from (6.37) that neglecting terms of the order of magnitude $\gamma_{cA}/w_{cc'}$, $\gamma_{cB}/w_{cc'}$ (they are small according to the condition (6.36)), it is possible to present P_c as

$$P_c=c_0e^{-\beta E_c}/\Sigma_c\equiv N_c\,e^{-\beta E_c} \qquad (6.39)$$

where c_0 is found by substituting (6.39) for (6.38):

$$c_0=\Sigma_c(\gamma_{CA}+\gamma_{CB})^{-1}(\gamma_{CA}C_A/\Sigma_A+\gamma_{CB}C_B/\Sigma_B) \qquad (6.40)$$

Here

$$\gamma_{CA}=\sum_c\gamma_{cA}e^{-\beta E_c}/\sum_c e^{-\beta E_c}$$

$$\gamma_{CB}=\sum_c\gamma_{cB}e^{-\beta E_c}/\sum_c e^{-\beta E_c} \qquad (6.41)$$

Substituting expressions for N_c, from (6.39), (6.40) for (6.23), (6.24), we again get the rate equation (6.32) with the rate constants equal

$$k_{AB}=\gamma_{CA}\gamma_{CB}(\gamma_{CA}+\gamma_{CB})^{-1}\exp[-\beta(F_C-F_A)]$$

$$k_{BA}=\gamma_{CA}\gamma_{CB}(\gamma_{CA}+\gamma_{CB})^{-1}\exp[-\beta(F_C-F_B)] \qquad (6.42)$$

The expressions (6.41), (6.42) coincide with those found by Gibbs and Fleming [59]. As we see their expressions are obtained as a specific case from the general formalism. Of course the expressions (6.91) satisfy the relation (6.35).

6.3 General features of adiabatic rate processes

Here, we summarize the main results of the theory of adiabatic rate processes, which has been developed in this chapter. A very general model of a nuclear subsystem having two equilibria positions and embedded in a condensed medium has been considered. This model, presented by the Hamiltonian (6.1), may be appropriate for the description of adiabatic rate processes such as proton transfer reactions, molecular group transfer, enzymatic reactions, etc. This model is also appropriate for description of adiabatic electron transfer reactions, accompanied by the molecular group transfer.

It has been shown that at temperatures (6.2): $k_BT \ll \hbar\omega_o$ (where $\hbar\omega_o$ is the characteristic energy difference in a nuclear subsystem) the model is isomorphous to the well explored model with two intersecting electronic energy hypersurfaces. Chapters IV and V were devoted to the investigation of this model. The implications for the adiabatic rate processes are the following.

Time behavior of adiabatic rate processes is described by the equations (5.26), (5.27) for 2x2 density matrix σ_{ik}, where its diagonal matrix elements σ_{11} and σ_{22} coincide with populations of state 1 in the well 1 and state 2 in the well 2 respectively. When the energy gap $(J_1 - J_2)$ between states 1 and 2 (where J_1 and J_2 are "dressed" energies (2.48) of the states 1 and 2) satisfies relation (5.43), the master equations for the populations σ_{11} and σ_{22} may be derived

$$\dot{\sigma}_{11} = -\dot{\sigma}_{22} = -\left(w_{12}\sigma_{11} - w_{21}\sigma_{22}\right) \tag{6.43}$$

The dependence of the transition rates w_{ik} on the temperature, the energy gap and other parameters has been investigated in Chapter IV.

In the case of symmetrical potential wells ($\omega_{12} = 0$) (and symmetrical surroundings), the transitions between wells (both adiabatic and non-adiabatic) are described not by the master equations, but by more complicated equations (4.104), (4.105), (5.59), (5.68) containing several parameters $w_{12} = w_{21}$, T_2^{-1}, T_3^{-1}, $\bar{V}/\hbar$.

The motion has mixed oscillatory and decaying character. At the low temperatures

$$k_BT \ll \hbar a = \hbar\omega_D$$

the probability of finding the particle in well 2 if it was initially in well 1 is determined by the parameter $\bar{V}/\hbar$ and not by the transition probabilities $w_{12} = w_{21}$. We have investigated the influence of the medium on the tunneling and quantum beats between two wells (or between two equivalent and symmetric states of the molecule embedded in the medium). This influence depends on strength of coupling between the molecule (nuclear subsystem) and the medium. The strength of coupling is characterized by the parameter $E_r/\hbar\omega_D$, where E_r (4.6) is the reorganization energy, and $\hbar\omega_D$ - is the Debye frequency.

An interesting situation arises in the case of strong coupling

$$E_r/\hbar\omega_D \gg 1 \tag{6.44}$$

In this case, even at zero temperature T = 0, quantum beats are strongly suppressed by the zero-point fluctuations of phonons (while the decay of these beats is determined by the spontaneous emission of phonons (5.72)). The frequency of quantum beats which is for a free molecule, is equal (2.19)

$$2V/\hbar$$

in the medium it becomes (5.27), (5.25), (4.12)) exponentially small

$$\Omega_0 = 2\bar{V}/\hbar = (2V/\hbar)\exp\left(-\tfrac{3}{4}E_r/\hbar\omega_D\right)$$

It should be stressed that, while the transition rates at low temperatures are mainly determined by the acoustic phonons, the "dressed" frequency Ω_0 (see (5.25)) is determined by all frequency ranges even at T = 0. Thus, it may happen that the Debye-Waller factor in (5.25) would be mainly determined by optical phonons at all temperatures, if the coupling with optical phonons is large enough. Of course the same refers to non-adiabatic processes. Starting from absolute zero temperature, two processes are competing: quantum beats with frequency Ω_0 and irreversible rate processes characterized by the quantity w (4.106), or γ(5.60). At the temperatures (4.138)

$$k_B T > \hbar\omega_D$$

the quantum beats are entirely suppressed; (4.143), and the rate process is becoming temperature activated and characterized by the activation energy (4.139) $E_r/4$. In the general case of now-symmetric potential wells the activation energy has the form (4.50)

$$E_a = (J_1 - J_2 + E_r)^2/4E_r$$

Here we have come to the important conclusion: three conditions

$$k_B T \ll \hbar\omega_0\,, \quad k_B T \gg \hbar\omega_D\,, \quad E_r \gg \hbar\omega_D \tag{6.45}$$

being satisfied, the transitions between the wells is thermally activated and characterized by the activation energy (4.50). This activation energy is determined by the properties of the medium and not by the barrier between two wells. This fact may be important for the interpretation of experimental data. It has a simple explanation. Two virtual reaction paths are competing. One is the overcoming of the barrier by the thermal fluctuations induced by the media, with activation energy U_a, where U_a is the height of the barrier. Another reaction path is characterized by medium coordinates and corresponds not to one "reaction coordinate" but to the continuum of the reaction coordinates.

In the case of weak coupling

$$E_r/\hbar\omega_D \ll 1$$

the influence of the medium is not so drastic as in the case (6.44). At low temperatures the frequency of the beats is almost unaffected. But in high temperature regions the quantum beats are also suppressed and the transition between two wells is temperature activated and described by the same formula (4.50).

All these considerations are valid for the temperature region

$$k_B T < \hbar\omega_0$$

In the case of strong coupling, this condition is compatible with (6.45) if

$$\omega_D < \omega_0$$

In the weak coupling case (with zero energy gap $J_1 - J_2 = 0$) we get more strict conditions (4.61)

$$\omega_0 \gg \omega_D (\hbar\omega_D/E_r) \gg \omega_D$$

At very high temperatures

$$k_B T \gg \hbar\omega_0$$

the transition rates again obey the Arrhenius law (see Section 6.2) but with the activation energy U_a, determined by the height of the barrier between two potential wells.

VII. COMPETITION BETWEEN ELECTRONIC AND VIBRATIONAL RELAXATIONS

7.1 Master equations describing competition between electronic and vibrational relaxations

One of the basic assumptions we have used in deriving the equations for the density matrix of dynamic subsystem was assumption (3.96). The meaning of this assumption is that relaxation processes in the dissipative system are going on much faster than those of the dynamic subsystem. In particular, it may mean that electronic relaxation is going on much more slowly than that of vibrational modes of dissipative system. It refers, e.g. to electron transfer processes in condensed media. However, slow vibrational relaxation has been recently reported [51] in the experiments connected with electronic excitation processes. Theory of coupled electronic-vibrational relaxation was developed by Jortner [52] and applied for the analysis of the luminescence of electronically excited centres in condensed phases. Kenkre [53] has analyzed the situations of slow vibrational relaxation using the master equation approach.

The question arises as to when use of master equations for such processes may be justified. To answer this question [36] we would generalize the model of electron transfer (Chapter 2.4), (or isomorphous to it, model of energy transfer (Chapter 2.5), or the model of adiabatic transition (Chapter 6.1)). We would single out "slow" vibrational modes, considering them as a part of the dynamic system. Mainly for simplicity's sake, we will consider only one singled out vibrational mode. But most of the conclusions do not depend on this assumption.

Thus, instead of the Hamiltonian (4.1) we will use the following Hamiltonian

$$\begin{aligned}\mathcal{H} = {} & n_1 E_1 + n_2 E_2 + \frac{1}{2}\left(P^2 + \omega_0 Q^2\right) + \frac{1}{2}\sum_k{}' \left(p_k^2 + \omega_k^2 q_k^2\right) \\ & - n_1\left(\omega_0^2 Q_1 Q + \sum_k{}' \omega_k^2 q_{1k}^{o2} q_k\right) - n_2\left(\omega_0^2 Q_2 Q + \sum_k{}' \omega_k^2 q_{2k}^{o} q_k\right) \\ & + V_{12} r_+ + V_{21} r_- + G(q_1, q_2 \ldots q_N) + W(Q, q_1, \ldots, q_N)\end{aligned} \tag{7.1}$$

Here Q - the coordinate (coordinates) of the singled out mode and W is its interaction energy with all other modes.

Making transformations similar to those in Chapter 4.1 (and using the same approximations) we get for transformed Hamiltonian, the expression

$$\mathcal{H} = n_1 J_1 + n_2 J_2 + \frac{1}{2}\left(P^2 + \omega_0^2 Q^2\right) + \frac{1}{2}\sum_k{}' \left(p_k^2 + \omega_k^2 q_k^2\right) +$$

$$+ r_+ \Pi^+ V_{12} + r_- \Pi V_{21} + G + W \qquad (7.2)$$

where

$$\Pi = e^{-\eta_0 (a^+ - a)} \prod_k \exp\left[-\eta_k (a_k^+ - a_k)\right]$$

$$\eta_0 = (\omega_0/2\hbar)^{1/2}(Q_1 - Q_2)\ , \quad \eta_1 = (\omega_k/2\hbar)^{1/2}(q_{1k}^0 - q_{2k}^0), \qquad (7.3)$$

a, a^+ are the operators of the annihilation and creation corresponding to Q; a_k, a_k^+ are those of q_k;

$$J_i = E_i - \frac{1}{2}\sum_k{}' \omega_k^2 q_{ik}^{o2} - \frac{1}{2}\omega_0^2 Q_i^2 \qquad (7.4)$$

First three terms of the r.h.s. of (7.2) represent the dynamic system,

$$E = n_1 J_1 + n_2 J_2 + \frac{1}{2}\left(P^2 + \omega_0^2 Q^2\right) \qquad (7.5)$$

the fourth term represents the dissipative system, the last term represents the perturbation energy.

We will restrict ourselves by the Markovian approximation. Then, according to (3.59), (3.60) and (7.2) we get

$$\dot{\sigma} = -i\hbar^{-1}\left[E + \bar{V}^{ef}, \sigma\right] - R^{VV}\sigma - R^{WW}\sigma - R^{VW}\sigma \qquad (7.6)$$

where $\bar{V}^{ef}$ is determined by the equation (3.60), R^{VV} is determined by (3.6) and (3.55), R^{WW} may be obtained from R^{VV} by substitution of W instead of V, and R^{VW} is defined by the expressions

$$\left(R^{VW}\right)_{mn}^{m'n'} = \int_0^\infty \left(\bar{K}^{VW}\right)_{mn}^{m'n'} e^{i\omega_{m'n'}\tau}\, d\tau \qquad (7.7)$$

$$\hbar^2 \left(\bar{K}^{VW}\right)^{m'n'}_{mn} = \delta_{nn'} \sum_{m_1} \langle \tilde{V}_{mm_1}(\tau)\, \tilde{W}_{m_1 m'}(0) + \tilde{W}_{mm_1}(\tau)\, \tilde{V}_{m_1 m'}(0) \rangle e^{-i\omega_{mn}\tau}$$

$$+ \delta_{mm'} \sum_{m_1} \langle \tilde{V}_{n'm_1}(0)\, W_{m_1 n}(\tau) + \tilde{W}_{n'm_1}(0)\, \hat{V}_{m_1 n}(\tau) \rangle e^{-i\omega_{mm_1}\tau} \tag{7.8}$$

$$- \langle \tilde{V}_{n'n}(0)\, \tilde{W}_{mm'}(\tau) + \tilde{W}_{n'n}(0)\, V_{mm'}(\tau) \rangle e^{-i\omega_{m'n}\tau}$$

$$- \langle \tilde{V}_{n'n}(\tau)\, \tilde{W}_{mm'}(0) + \tilde{W}_{n'n}(\tau)\, \tilde{V}_{mm'}(0) \rangle e^{-i\omega_{mn'}\tau}$$

The Hamiltonian of the dynamic system (7.5) has the eigenvalues

$$E_{jl} = J_j + \left(l + \tfrac{1}{2}\right)\hbar\omega_0 , \quad j = 1,2 ; \; l = 0,1,2,\ldots \tag{7.9}$$

We will consider two extreme cases:

$$J_1 - J_2 = k\hbar\omega_0 \tag{7.10}$$

and

$$\left| J_1 - J_2 - k\hbar\omega_0 \right| \gg \left| \bar{V}^{ef} \right| / \hbar , \; T^{-1}_{dyn} \tag{7.11}$$

where k - the integer number (positive or negative), and T_{dyn} is the characterisitc time of the dynamic system.

First we will consider the case (7.11). In this case all the levels of the dynamic system are non-degenerate and conditions for the validity of the master equations are satisfied (see Chapter 3.3). The inspection of the expressions (7.7), (7.8) shows that

$$\left(R^{VW}\right)^{mm}_{nn} = 0 \tag{7.12}$$

Thus, according to (3.68), (7.9), (7.12) the master equations describing the relaxation of the dynamic system have the form

$$\dot{P}_{jl} = -\sum_{j' \neq j, l'} \left(w_{jl,j'l'} P_{jl} - w_{j'l',jl} P_{j'l'} \right) - \tag{7.13}$$
$$- \sum_{l'} \left(D_{ll'} P_{jl} - D_{l'l} P_{jl'} \right)$$

where $P_{jl} = \sigma_{jl,jl}$ are the probabilities of finding the system in the state j,l; j - electronic quantum number, l - vibrational quantum number.

This equation (7.13) coincides with that used by Kenkre [52] (see formula (1.5) of his paper), provided the condition

$$w_{jl,j'l'} = w_l^{jj'} \delta_{ll'} \tag{7.14}$$

is satisfied. The justification of such a condition may follow from the analysis of certain concrete systems. In a general case, it is not satisfied. It is worthwhile mentioning that in derivation of the master equation (7.13) the fact that we have singled out only one vibrational mode is not essential.

For concrete Hamiltonian (7.2), using expressions (3.69), (4.28) we get expressions for the transition probabilities

$$W_{jl,j'l'} = V^2 \hbar^{-2} \left| \langle l | e^{-\gamma_0 (a^+ - a)} | l' \rangle \right|^2 e^{-\int_{-a}^{a} \rho(\omega) d\omega} \times$$
$$\times \int_{-\infty}^{\infty} \left\{ \exp\left[\int_{-a}^{a} \rho(\omega) e^{-i\omega\tau} d\omega \right] - 1 \right\} \exp\left[i\omega_{jj'}\tau + i\omega_0 (l-l')\tau \right] d\tau \tag{7.15}$$

The transition probabilities $D_{ll'}$ have the form

$$D_{ll'} = 2\pi\hbar^{-1} \sum_{\alpha,\alpha' \, (\alpha \neq \alpha')} P_\alpha \left| W_{l\alpha,l'\alpha'} \right|^2 \delta\left[F_\alpha - F_{\alpha'} - (l-l')\hbar\omega_0 \right] \tag{7.16}$$

They determine the transitions among various vibrational levels without change of the electronic state. In the case of fast vibrational relaxation

$$D_{ll'} \gg w_{jl,j'l'} \tag{7.17}$$

the equations (7.13), (7.15) may be reduced to the expressions employed in literature [54] for the model with one singled out vibrational mode. In the case (7.17) using the same argument as in Chapter 3.2, (3.46)-(3.52) one can show that with accuracy $w_{jl;jl'}/D_{ll'}$ the density matrix may be factorized

$$\sigma_{jl,j'l} = \sigma_{jj}\, e^{-l\hbar\omega_0} \Big/ \sum_k{}' e^{-k\hbar\omega_0} \tag{7.18}$$

and the equation for the electronic concentrations takes the form of simple two-level master equations

$$\dot{C}_1 = -\dot{C}_2 = -w_{12} C_1 + w_{21} C_2 \tag{7.19}$$

where

$$w_{jj'} = \sum_{ll'}{}' P_l w_{jl,j'l'} = V^2 \hbar^{-2} \exp\left[-\int_{-a}^{a} \rho(\omega)\, d\omega\right] \times$$

$$\times \int_{-\infty}^{\infty} d\tau \left[e^{\int_{-a}^{a} \rho(\omega) e^{-i\omega\tau} d\omega} - 1 \right] \sum_{ll'} P_l \left| \langle l | e^{-\xi_0 (a^+ - a)} | l' \rangle \right|^2 e^{i\omega_0 (l-l')\tau}\, e^{i\omega_{jj'}\tau}$$

$$C_1 = \sigma_{11}\,, \qquad C_2 = \sigma_{22}$$

After the transformations we get

$$w_{jj'} = V^2 \hbar^{-2} e^{-\int_{-a}^{a} \rho(\omega)\, d\omega} \int_{-\infty}^{\infty} \left[e^{\int_{-a}^{a} \rho(\omega) e^{-i\omega\tau} d\omega} - 1 \right] \times$$

$$\times \exp\left\{ \xi_0^2 \left[i \sin \omega_0 \tau - (2n+1)(1 - \cos \omega_0 \tau) \right] \right\} e^{i\omega_{jj'}\tau}\, d\tau \tag{7.20}$$

$$= V^2 \hbar^{-2} e^{-\int_{-a}^{a} \rho(\omega)\, d\omega}\, e^{-\xi_0^2 (2n+1)} \int_{-\infty}^{\infty} \left[e^{\int_{-a}^{a} \rho(\omega) e^{-i\omega\tau} d\omega} - 1 \right] \times$$

$$\times \sum_k{}' \left((n+1)/n \right)^{k/2} I_k \left[2\xi_0 \left(n(n+1) \right)^{\frac{1}{2}} \right] e^{i\omega_{jj'}\tau + ik\omega_0\tau}\, d\tau$$

where n is the thermal equilibrium average number of phonons with frequency ω_0 and $I_k(z)$ is the modified Bessel function. Thus we have come to the expression obtained in [53]. The term -1 in the curly brackets does not play any role here; due to the assumption (7.11) the exponent $\omega_{jj'} + k\omega_0$ can never reach zero.

Now, we will briefly examine the case (7.10) which corresponds to the degenerate energy levels of the dynamic subsystem. For simplicity, we will put k = 0 in (7.10). This is the case of the symmetric electronic hypersurfaces. In this case all levels of the system have double degeneracy

$$E_{1\ell} = J + (\ell + \tfrac{1}{2})\hbar\omega_0 = E_{2\ell}\,, \quad J_1 = J_2 = J \tag{7.21}$$

Now the equation (7.6) for the density matrix of the dynamic subsystem cannot be reduced to master equations and apart from highly oscillating terms (which are small if

$$\omega_0 \gg |\bar{V}^{ef}|/\hbar\,, \; T_{dyn}^{-1}$$

has the form

$$\dot{\sigma}_{1\ell,1\ell} = -i\hbar^{-1}\left(\bar{V}^{ef}_{1\ell}\,\sigma_{2\ell,1\ell} - \sigma_{1\ell,2\ell}\,\bar{V}^{ef}_{2\ell,1\ell}\right) -$$

$$- \sum_{\ell'}{}' \left(w_{1\ell,2\ell'}\,\sigma_{1\ell,1\ell} - w_{2\ell',1\ell}\,\sigma_{2\ell',2\ell'}\right) -$$

$$- \sum_{\ell'}{}' \left(D_{\ell\ell'}\,\sigma_{1\ell,1\ell} - D_{\ell'\ell}\,\sigma_{1\ell',1\ell'}\right) -$$

$$- \sum_{\ell'}{}' \left[(R^{VW})^{1\ell'2\ell'}_{1\ell,1\ell}\,\sigma_{1\ell',2\ell'} + (R^{VW})^{2\ell',1\ell'}_{1\ell,1\ell}\,\sigma_{2\ell',1\ell'} \right]$$

Similar equations may be written for $\sigma_{2\ell 2\ell}$, $\sigma_{2\ell 1\ell}$, $\sigma_{1\ell,2\ell}$. Here $(R^{VW})^{1\ell 2\ell'}_{1\ell 1\ell}$, $(R^{VW})^{2\ell' 1\ell'}_{1\ell 1\ell}$ are defined by (7.7), (7.8). The equations for $\sigma_{1\ell 1\ell}$, $\sigma_{2\ell 2\ell}$, $\sigma_{2\ell 1\ell}$, $\sigma_{1\ell 2\ell}$ do not coincide with master equations and describe mixed oscillatory decaying processes in the dynamic subsystem.

7.2 Master equation approach to non-adiabatic rate processes

The results of the preceding section may be generalized, if we postulate the validity of the master equation of (7.13) type, without the reference to the specific model. Using this approach we can develop a general theory of non-adiabatic rate processes.

We are dealing with two intersecting potential energy hypersurfaces: A and B (Fig. 10). The minimum of each potential energy hypersurface corresponds to (quasi)

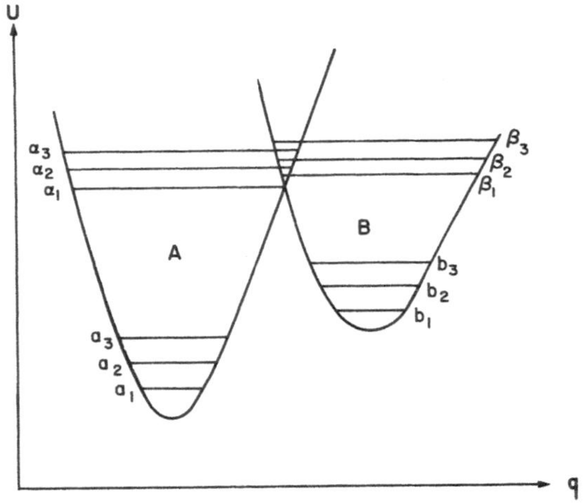

Fig. 10. Potential energy curves describing non-adiabatic rate processes.

stable configuration of nuclei. The experience of the theory of Landau-Zener transitions (see Chapter 2.3) teaches us that at high enough temperatures (neglecting the tunneling) the effective transitions are between microstates a and b (of A and B) which are lying above the intersection point with minimal energy.

Thus we introduce two kinds of microstates A: a and α, and two kinds of microstates B: b and β. Microstates α and β lie above the point of intersection with minimal energy. The transition probabilities between these states are assumed to be much larger than those among states a and b (the latter correspond to the tunneling).

The corresponding master equations take the form

$$\dot{P}_a = -\sum_{a'}{}' (w_{aa'} P_a - w_{a'a} P_{a'}) - \sum_{\alpha}{}' (w_{a\alpha} P_a - w_{\alpha a} P_\alpha)$$

$$\dot{P}_b = -\sum_{b'}{}' (w_{bb'} P_b - w_{b'b} P_{b'}) - \sum_{\beta}{}' (w_{b\beta} P_b - w_{\beta b} P_\beta)$$

$$\dot{P}_\alpha = -\sum_{a} (w_{\alpha a} P_\alpha - w_{a\alpha} P_a) - \sum_{\beta}{}' (w_{\alpha\beta} P_\alpha - w_{\beta\alpha} P_\beta) - \sum_{\alpha'} (w_{\alpha\alpha'} P_\alpha - w_{\alpha'\alpha} P_{\alpha'})$$

$$P_\beta = -\sum_{\beta}{}' (w_{\beta b} P_\beta - w_{b\beta} P_b) - \sum_{\alpha}{}' (w_{\beta\alpha} P_\beta - w_{\alpha\beta} P_\alpha) - \sum_{\beta'}{}' (w_{\beta\beta'} P_\beta - w_{\beta'\beta} P_{\beta'})$$

(7.22)

First let us consider the case when the transition probabilities between different electronic states are much smaller than those among different levels at the same state A or B

$$w_{\alpha\beta} << w_{aa'}, w_{\alpha\alpha'}, w_{\alpha a}, w_{bb'}, \ldots \tag{7.23}$$

In this case the probability distribution of microstates of state A or B may be approximated by the expressions

$$\begin{aligned} P_a &= C_A(t) e^{-\beta E_a} / \Sigma_A' + O(w_{\alpha\beta}/w_{aa'}, \ldots) \\ P_\alpha &= C_A(t) e^{-\beta E_\alpha} / \Sigma_A' + O(w_{\alpha\beta}/w_{\alpha\alpha'}, \ldots), \ etc. \end{aligned} \tag{7.24}$$

where

$$\Sigma_A' = \sum_a' e^{-\beta E_a} + \sum_\alpha' e^{-\beta E_\alpha} \tag{7.25}$$

Here we neglect corrections to the thermodynamic equilibrium in each state of A and B. These corrections are small due to the condition (7.23).

Substituting expressions (7.24) (without small corrections) into the master equations (7.22) we get the rate equations for the overall concentrations C_A and C_B of species A and B

$$\dot{C}_A = -k_{AB} C_A + k_{BA} C_B \tag{7.26}$$

where the rate constants are equal to

$$\begin{aligned} k_{AB} &= \sum_{\alpha\beta}' w_{\alpha\beta} e^{-\beta E_\alpha} / \Sigma_A' \\ k_{BA} &= \sum_{\alpha\beta}' w_{\beta\alpha} e^{-\beta E_\beta} / \Sigma_B' \end{aligned} \tag{7.27}$$

These expressions are valid also in the case when α, β take all the meanings, including a and b. It means that they may also take into account the tunneling between the microstates of A and B. These expressions may be used at all temperatures.

In the case of high temperatures when the tunneling may be neglected and at the same time at the temperatures satisfying the condition (6.19), the expression may be approximately rewritten in the form explicitly exhibiting the Arrhenius law

$$k_{AB} = w \exp\left(-\beta\left(\tilde{F}_A - F_A\right)\right), \quad k_{BA} = w \exp\left(-\beta\left(\tilde{F}_B - F_B\right)\right) \tag{7.28}$$

where

$$w = \sum_{\alpha\beta}' w_{\alpha\beta} e^{-\beta E_\alpha} \Big/ \sum_{\alpha}' e^{-\beta E_\alpha} \tag{7.29}$$

$$e^{-\beta \tilde{F}_A} = \sum_{\alpha}' e^{-\beta E_\alpha}, \quad e^{-\beta F_A} = \sum_{a}' e^{-\beta E_a} \tag{7.30}$$

It is easy to check that the thermodynamic equilibrium relation (6.35) is satisfied by the expressions (7.27)-(7.30).

Now, let us consider the case opposite to (7.23), i.e. the case

$$w_{\alpha\beta} \gg w_{\alpha\alpha'}, \; w_{\beta\beta'} \tag{7.31}$$

We assume also that the condition (6.19) is satisfied.

This case may be analyzed similarly to the case (6.36) of adiabatic transitions. The resulting rate constants have the form

$$\begin{aligned} k_{AB} &= \gamma_A \gamma_B \left(\gamma_A + \gamma_B\right)^{-1} e^{-\beta\left(F_0 - F_A\right)} \\ k_{BA} &= \gamma_A \gamma_B \left(\gamma_A + \gamma_B\right)^{-1} e^{-\beta\left(F_0 - F_B\right)} \end{aligned} \tag{7.32}$$

where the free energy of the "activated complex" (of the "transition state") is defined as

$$e^{-\beta F_0} = \sum_{\alpha}' e^{-\beta E_\alpha} + \sum_{\beta}' e^{-\beta E_\beta} = \sum\nolimits_0' \tag{7.33}$$

and

$$e^{-\beta F_A} = \sum_a e^{-\beta E_a} \; , \; e^{-\beta F_B} = \sum_b e^{-\beta E_b} \qquad (7.34)$$

The constants γ_A, γ_B are defined as

$$\gamma_A = \sum_{\alpha a} w_{\alpha a} e^{-\beta E_\alpha} / \sum_0 \; , \; \gamma_B = \sum_{\beta b} w_{\beta b} e^{-\beta E_\beta} / \sum_0 \qquad (7.35)$$

As it is seen from the comparison of the formulae (6.42) and (7.38), the non-adiabatic transitions in the case (7.31) exhibit very close similarity to the adiabatic transitions. In this case the reaction rates k_{AB}, k_{BA} are determined not by transition probabilities $w_{\alpha\beta}$ between states A and B, but by the relaxation rates $w_{\alpha a}$, $w_{\beta b}$ between various microstates at the same electronic hypersurfaces A or B.

VIII. CONCLUDING REMARKS

This work has been aimed at reviewing and analyzing the theoretical aspects of rate processes in condensed media. The consideration has been undertaken at three different levels: (1) general quantum mechanical, (2) derivation and analysis of equations of motion, and (3) concrete models and calculation of transition rates.

Here we present some implications of these considerations.

8.1 General quantum mechanical analysis

One of the conclusions of the general quantum mechanical analysis (Chapter I) was the statement that the irreversibility of rate processes is connected with a continuous spectrum of energies. There are three kinds of systems having continuous energy spectra.

(1) A continuous energy spectrum may be provided by infinite, unbounded motion. Thus free particles with definite kinetic energy are characterized by continuous energy spectrum. Of course the same refers to finite systems (with boundaries), when a periodic motion takes place, but with a very large period. An example of such a situation is shown in Fig. 4 (page 39). In this case we are dealing with quasi-continuous spectrum of energies. Let $\hbar\omega_0$ be the energy difference between adjacent energy levels. The energy spectrum may be considered as a continuous one if the time interval

$$T_0 = \omega_0^{-1}$$

is larger than all characteristic times of the system (including the duration of the experiment). The case of infinite (or quasi-infinite) motion corresponds to the rate processes in gaseous media. Particularly it refers to the chemical reactions in gaseous phase, where the reaction proceeds due to the kinetic energy of reaction components.

(2) The second class of systems which may exhibit the irreversibility are nonlinear semiclassical systems performing finite motion. As is known [2], such systems are characterized by the energy difference between adjacent energy levels

$$\Delta E = E_{n+1} - E_n = \hbar\omega(n) \tag{8.1}$$

where ω is the frequency of classical motion. In semiclassical approximation, $\omega(n)$ depends on n very slowly. In this case the action variable is equal

$$J = n\hbar$$

and

$$\frac{\partial\omega}{\partial n} = \frac{\partial\omega}{\partial J}\frac{\partial J}{\partial n} = \frac{\partial\omega}{\partial J}\hbar \tag{8.2}$$

This derivative tends to zero when $\hbar \to 0$. The mean value of some physical quantity may be written in the semiclassical approximation as [2]

$$\langle A(t)\rangle = \sum_{n,s} \rho_{n,n+s} A_{n+s,n} e^{is\omega(n)t}$$

where $n \gg 1$; $n\hbar$ is a finite quantity, and $s \ll n$.

Assuming $\rho_{n,n+s}$ to be a smooth function of s we get from (8.2) the semiclassical expression for the Fourier expansion of the function

$$\langle A(t)\rangle = \sum_{n} \rho_{nn} A_s(n) e^{is\omega(n)t} \tag{8.3}$$

where $A_s(n) = A_{n+s,n}$.

If the system is in the state with definite energy (with definite quantum number n), the expression (8.3) describes periodical motion with the period

$$T = 2\pi/\omega(n)$$

Thus in classical limit and considering only one trajectory we have come to the natural conclusion that closed bounded system performs periodical motion and does not exhibit any irreversibility. The situation changes when we consider a bunch of trajectories characterized by finite energy interval

$$\Delta E = \Delta n \hbar\omega(n)$$

with

$$1 \ll \Delta n \ll n$$

and finite

$$\hbar\Delta n, \quad \text{when } \hbar \to 0$$

In this case the summation over n in (8.3) may be replaced by the integration over E and we are dealing with a Fourier integral over frequencies ω. The quantity $\langle A(t)\rangle$ would tend to a definite limit (cf. (1.86)) in time

$$t \gg \tau = 1\Big/\frac{\partial\omega(J)}{\partial J}\Delta J \tag{8.4}$$

where ΔJ corresponds to the energy interval ΔE.

Thus the bounded system with one or several degrees of freedom exhibit the irreversibility, provided the following conditions are fulfilled.

(a) The system is semiclassical

$$n \gg 1$$

(b) The system is non-linear

$$\frac{\partial\omega}{\partial J} \neq 0$$

(c) The irreversibility is connected with a finite volume ΔJ in the phase space (8.4)

$$\frac{\partial \omega}{\partial J} \Delta J t \gg 1$$

In particular, this interval ΔJ may be determined by the accuracy of the measuring device.

All these three conditions are necessary conditions of the irreversibility displayed by finite bounded systems with one or several degrees of freedom.

It is easy to check this statement. If condition (a) is violated, then we are dealing with several discrete energy levels and frequencies. It is clear that such a system would not exhibit any irreversibility. When condition (b) is violated and the system is linear, ω does not depend on n. The sum (8.3) would be a periodic function of time. And last, when condition (c) is violated and only several quantum numbers n contribute to the r.h.s. of (8.3),we are again dealing with a quasi-periodic but reversible system.

The investigation of the irreversibility of bounded classical nonlinear systems is now a developing field of research [55-57]. It may have physical implications for non-radiative processes in molecules.

(3) Third class of processes which may exhibit the irreversibility are those in condensed media.

Typical for rate processes in condensed media is the existence of a dynamic subsystem interacting with its surroundings. The dynamic system itself - without a condensed medium - performs reversible motion. It is bounded and, generally speaking, quantum. Thus the irreversibility is connected with the condensed medium. Each mode of motion of the condensed medium, e.g. a vibrational mode, also performs a reversible motion. But, for large enough volume of a condensed medium,the frequencies of its modes form a quasicontinuous spectrum. For example, in a periodic structure the wave vectors have quantized values

$$k_x = 2\pi n_x / A \; , \quad k_y = 2\pi n_y / B \; , \quad k_z = 2\pi n_z / C$$

where A, B, C are the dimensions of the volume and n_x, n_y, n_z are integer numbers. The eigen-frequencies of the modes of the condensed medium are functions of these wavevectors

$$\omega = \omega(\vec{k})$$

and thus they form quasicontinuous spectra, if the dimensions A, B, C are large enough.

Therefore,in the case of rate processes in condensed media,the continuous energy spectrum, and thus the irreversibility, is provided by a continuum of degrees of freedom of the condensed medium. This is the characteristic feature of rate processes in condensed media.

8.2 Hierarchy of equations of motion

Now, we are going to summarize conclusions in a more specific way, but still at very general level of consideration - the general framework of equations describing rate processes in condensed media.

The hierarchy of the equations of motion describing various rate processes in condensed media is presented in Table 1. In this diagram solid arrows indicate exact consequences from corresponding equations, the wave arrows represent the approximate results.

We begin with the Hamiltonian (3.2)

$$\mathcal{H} = E + F + V + G$$

and the von Neumann equation (1.23), (3.1). For rate processes in condensed media, E represents a dynamic subsystem, F - a condensed medium, V - the interaction energy between the subsystem and the condensed medium; G is the perturbation energy responsible for the relaxation processes in the condensed medium itself. The von Neumann equation (3.1) with the Hamiltonian (3.2) is the starting point of the consideration. The derivation of approximate equations, appropriate for description of rate processes in condensed media, is connected with relations between various time scales.

We are mainly interested in a set of blocks descending from the central one with the initial von Neumann equation (Table 1). But first let us consider some implications of the von Neumann equation with V = 0. In this case the density matrix of the whole system may be factorized

$$\rho_{m\alpha n\beta} = \sigma_{mn}\,\rho_{\alpha\beta} \tag{8.5}$$

where σ_{mn} is the density matrix of dynamic subsystem, and $\rho_{\alpha\beta}$ is that of the dissipative subsystem (a condensed medium). Two von Neumann equations (two upper blocks) are exact implications of the initial von Neumann equation. In this case (V=0), the dynamic subsystem performs reversible motions determined by its Hamiltonian E. Further simplification follows for G=0. The von Neumann equation of the dissipative system is reduced to the following one (third upper block, right side)

$$\dot{\rho}_{\alpha\beta} = -i\omega_{\alpha\beta}\,\rho_{\alpha\beta}\,, \quad \omega_{\alpha\beta} = (F_\alpha - F_\beta)/\hbar \tag{8.6}$$

The average meaning of some quantity A represented by the matrix elements $A_{\alpha\beta}$ reads

$$\langle A(t)\rangle = \sum_{\alpha\beta}{}' \rho_{\alpha\beta}(0)\,A_{\beta\alpha}\,e^{i\omega_{\beta\alpha}t} \tag{8.7}$$

Here we should utilize the assumption about continuous energy spectrum of a condensed medium. Then the sum (8.7) may be represented as the Fourier integral over the frequencies $\omega_{\alpha\beta}$. If ω^* is the characteristic interval of these frequencies, then the sum (8.7) would come to its asymptotic value in time larger than

Table 1

Hierarchy of equations of motion

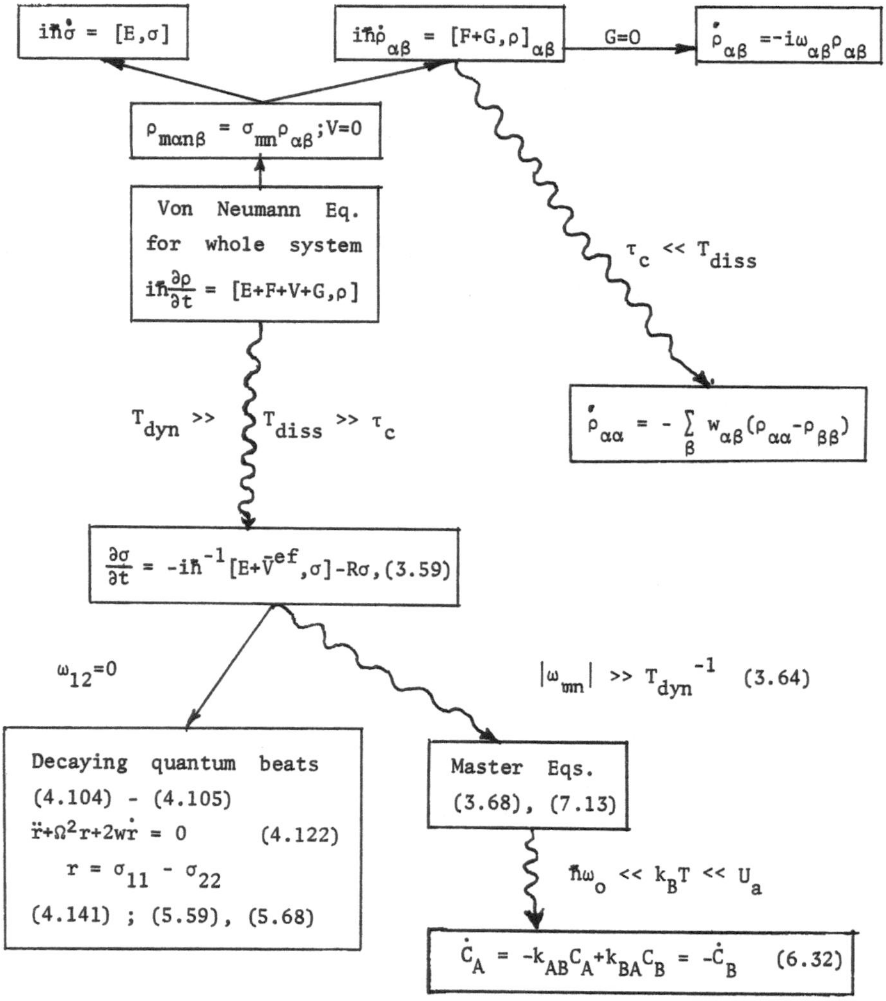

$$\tau_c = \omega^{*\,-1} \tag{8.8}$$

On the other hand diagonal elements of the density matrix $\rho_{\alpha\alpha}$, representing the probability to find the system in some state α do not depend on time at all. The same relates to the quantities B represented by diagonal matrixes $B_{\alpha\alpha}$.

Now, if the perturbation energy G is not zero but small enough, so that the condition

$$T_{diss}^{-1} = 2\pi\hbar^{-1}\sum_{\beta}{}' |G_{\alpha\beta}|^2 \delta(F_\alpha - F_\beta) \ll \omega^* = \tau_c^{-1} \tag{8.9}$$

is satisfied, we get for the diagonal elements $\rho_{\alpha\alpha}$ the master equation (3.32) (see Table 1). Of course, we are mainly interested in the region $V \neq 0$. We assume that V is small enough, so that approximate factorization (8.5), (3.49) still holds. It means that characteristic time T_{dyn} (determined by V^2, (3.69)) is large enough*

$$T_{dyn} \gg T_{diss} \gg \tau_c \tag{8.10}$$

In this case we perform approximate reduction to the equation (3.50), governing the time evolution of the density matrix σ of the dynamic system only. The first term in the r.h.s. of this equation describes dynamical motion and the second term is responsible for irreversible relaxation processes.

In particular, for two-state dynamic subsystem with equal energy levels ($\omega_{12}=0$) the equation is reduced to the equations (4.14), (4.15) or (4.122), (4.141). These equations describe decaying quantum oscillations between two states of the dynamic subsystem.

On the other hand for the dynamic system with non-degenerate levels satisfying condition (3.64), the equation (3.59) may be reduced to the master equations (3.68) (see Table 1).

If there are two groups of states a and b divided by the large enough potential barrier U_a, the master equations may be reduced to the rate equations for the concentrations (see section (6.2))

$$c_A = \sum_a{}' \sigma_{aa} \quad , \quad c_B = \sum_b{}' \sigma_{bb}$$

This reduction is possible if the condition (6.20)

$$\hbar\omega_0 \ll k_B T \ll U_\alpha$$

is satisfied.

---

*) Here we make the quite reasonable assumption that various correlation times τ_c of the dissipative system (3.29), (3.58) and others are of the same order of magnitude.

It is worth-while mentioning that comparatively simple master equations and reduction to the variables of the dynamic subsystem only, are results of the existence of small parameters $|G^2|$, $|V^2|$ and $|V^2|/|G^2|$. The corresponding small dimensionless parameters are

$$\lambda_1 = \tau_c / T_{diss}, \quad \lambda_2 = \tau_c / T_{dyn}, \quad \lambda_3 = T_{diss} / T_{dyn} \tag{8.11}$$

When these parameters are not small, comparatively simple master equations cannot be derived. For the case when λ_1, is not small, Van Hove [58] has shown that much more complicated equations with memory take place.

The equation (3.59) for the dynamic subsystem has been derived under the assumption that the relaxation processes in a condensed medium (a dissipative system) are going on much faster than those in the dynamic subsystem

$$\lambda_3 = T_{diss} / T_{dyn} \ll 1 \tag{8.12}$$

If this condition is not satisfied for some modes of motion (e.g. for some vibrational modes) of the dissipative system, then these modes* may be included in the dynamic subsystem. In this case we are dealing with the competition between electronic and vibrational relaxations. In the case of non-degenerate levels of the dynamic system, this competition may be described by the master equations (7.13).

It is worth-while to stress that till now the whole analysis is quite general and is not connected with particular models. Table 1 shows the relations between equations of motion describing various rate processes and under which conditions they may be derived from the first principles.

However, for concrete calculations of coefficients in these equations (transition rates) the use of models is necessary. And here we come to the third level of the consideration.

8.3 Models

Again we can establish a certain hierarchy (see Table 2). At the top of this hierarchy we have the Born-Oppenheimer approximation. In the framework of this approximation the rate processes may be considered as transitions between two (quasi)stable configurations of the nuclei. This statement is obvious as far as we consider chemical transformations, or nuclear group transfer, since they are nothing other than transitions between various (quasi)stable configurations of nuclei. But, as it is

---

*) Of course it is assumed that these modes give non-vanishing contribution to the relaxation process, when total number N of modes tends to infinity.

Table 2

Hierarchy of models and approximations

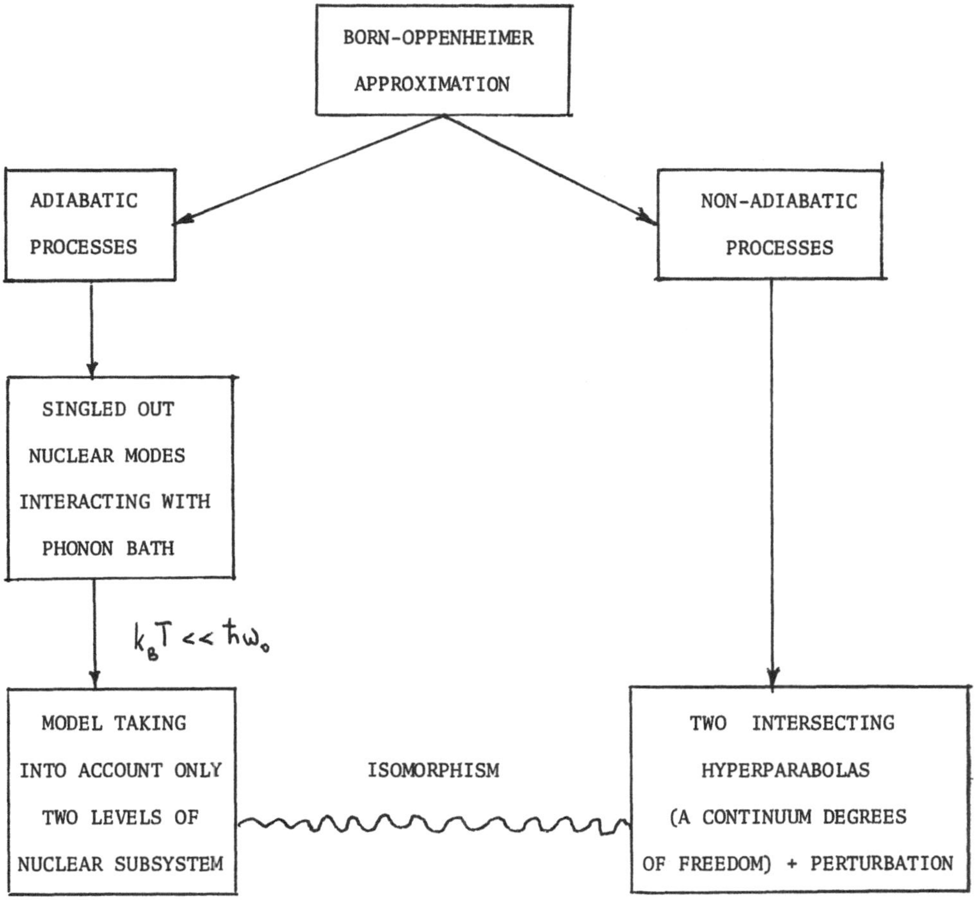

clear from the above (sections 2.4, 2.5), such processes as the electron transfer or energy transfer also may be described as transitions between two quasi(stable) configurations of nuclei.

There are two possibilities. Transitions may happen at the same electronic energy hyper-surface - adiabatic transitions. Another possibility is connected with transitions between various electronic energy hyperfurfaces - non-adiabatic transitions. In the latter case, in order to perform analytical calculations,the hypersurfaces are modeled by the intersecting hyperparabolas with the same curvatures (frequencies) but with different positions of minima (equilibria). Transitions between two equilibria positions are caused by the small perturbation energy term. In the case of adiabatic transitions (Chapter VI),another model has been worked out. In this model several degrees of freedom are singled out and they interact with a phonon bath, representing a condensed medium. Such a model may describe a transfer of molecular groups in condensed media. If the condition (6.2)

$$k_B T \ll \hbar\omega_0 \tag{8.13}$$

is fulfilled, i.e. if $k_B T$ is much smaller than the characteristic energy interval between the adjacent energy levels in the nuclear subsystem, then it has been shown (Chapter VI) that the adiabatic processes may be described by the model (6.8). This model is isomorphous to that accepted for description of non-adiabatic processes (two intersecting hyperparabolas).

Mentioning this isomorphism we conclude the presentation of the hierarchy of approximations and models (Table 2). What is lacking in the diagram and in the present work as a whole is the model of adiabatic transitions on one electronic hypersurface with a continuum of degrees of freedom and without singled out modes. In other words we cannot present a reasonable model which is in a sense an adiabatic counterpart of the above mentioned model with two intersecting hyperparabolas. In this case the characteristic energy difference between adjacent levels tends to zero when $N \to \infty$ (i.e. for a continuum of degrees of freedom) and the condition (6.2), (8.13) cannot be satisfied.

The essential part of the work (Chapter IV) has been devoted to calculations of transition rates in the model with two intersecting hypersurfaces. As we have seen this model cover a number of rate processes - both non-adiabatic and adiabatic ones.

We do not intend to review again the results of these calculations. But two remarks should be made. First, these calculations take into account a continuum of degrees of freedom of condensed media. Second, some of the obtained relations do not depend on the specificity of the model and have more general meaning. One example of such a relation is the Arrhenius law

$$w_{12} \propto e^{-E_a/k_B T}$$

The activation energy temperature dependence is a direct consequence of the assumption about fast relaxation in the dissipative system towards the Gibbs distribution. (On the other hand a pre-exponential factor essentially depends on the model). Another example of very universal dependence is the energy gap law (4.102)

$$w_{12} \propto \exp\left[-\frac{|J_1 - J_2|}{\hbar\omega_{max}} \ln\left(\frac{|J_1 - J_2|}{\hbar\omega_{max}}\right)\right]$$

For large enough energy gaps $|J_1 - J_2|$ this dependence follows from general formulae of non-stationary perturbation theory (Section 4.3).

Index

A

B

C

D

E

G

H

I

L

M

N

P

Q

R

S

T

U

V

W

REFERENCES

1) L. BRILLOUIN, Science and Information Theory, Academic Press, New York (1965).

2) L.D. LANDAU and E.M. LIFSHITZ, Quantum Mechanics, Pergamon Press, London (1958).

3) W. PAULI, Festschrift zum 60. Gebürstage A. Sommerfelds, Leipzig (1928).

4) S. GOLDEN and H.C. LONGUET-HIGGINS, J. Chem. Phys. 33, 1479 (1960).

5) B. FAIN and Y.I. KHANIN, Quantum Electronics, Pergamon, Oxford, New York (1969); B. FAIN, Photons and Non-Linear Media (in Russian) Moscow (1972).

6) M. BORN and KUN HUANG, Dynamical Theory of Crystal Lattices, Oxford (1954).

7) F. HUND, Z. Physik 43, 805 (1927).

8) G. GAMOV, Z. Physik 51, 204 (1928).

9) E.P. WIGNER, Z. Phys. Chem. (Leipzig) B19, 203 (1932).

10) S. GLASSTONE, K.Y. LAIDLER and H. EYRING, The Theory of Rate Processes, New York (1941).

11) H.A. KRAMERS, Physica VII, 284 (1940).

12) Y. CHANDRASEKHAR, Stochastic Problems in Physics and Astronomy, Rev. Mod. Phys. 15, 1-89 (1943).

13) Y. SKINNER and P. WOLYNESS, J. Chem. Phys. 69, 2143 (1978).

14) M.P. REAR and J.H. WEINER, J. Chem. Phys. 69, 15 (1978).

15) D. CHANDLER, J. Chem. Phys. 68, 2959 (1978).

16) J.H. WEINER, J. Chem. Phys. 68, 2492 (1978).

17) W.H. MILLER, Accounts Chem. Res. 9, 306 (1976).

18) C. BLOMBERG, Physica 86A, 49-66 (1977).

19) N.G. VAN KAMPEN, Journ. Statist. Phys. 17, 7 (1979).

20) R.A. MARCUS, J. Chem. Phys. 24, 966 (1956); Discuss. Faraday Soc. 29, 21 (1960).

21) V.G. LEVICH, Adv. Electrochem. Electrochem. Eng. 4, 249 (1956).

22) N.R. KESTNER, J. LOGAN and J. JORTNER, J. Phys. Chem. 78, 2148 (1974).

23) TH. FÖRSTER, Ann. Phys. (Leipzig) 2, 55 (1948).

24) D.L. DEXTER, J. Chem. Phys. 21 (1953).

25) G.W. ROBINSON and R.P. FROSCH, J. Chem. Phys. 38, 1187 (1963).

26) R. SILBEY, Ann. Rev. Phys. Chem. 27, 203 (1976).

27) V.M. KENKRE, Phys. Rev. B11, 1741 (1975); 12, 2150 (1975).

28) A.H. ZEWAIL and C.B. HARRIS, Phys. Rev. B11, 935, 11, 952 (1975).

29) TH. FÖRSTER, in Comparative Effects of Radiation, ed. by M. Burton, New York (1960).

30) A.S. DAVYDOV, Phys. Stat. Sol. 30, 357 (1968), D.L. DEXTER, TH. FÖRSTER, R.S. KNOX, Phys. Stat. Sol. 34, k 1959 (1969).

31) T.F. SOULES and C.B. DUKE, Phys. Rev. B, 3, 262 (1971).

32) S. RACKOVSKY and R. SILBEY, Mol. Phys. 25, 61 (1973).

33) I.I. ABRAM and R. SILBEY, J. Chem. Phys. 63, 2137 (1975).

34) L. VAN HOVE, Physica 21, 517 (1955).

35) R.W. ZWANZIG, J. Chem. Phys. 33, 1338 (1960).

36) B. FAIN, Physica A, 101A, 67 (1980).

37) F. BLOCH, Phys. Rev. 105, 1206 (1957).

38) P.S. HUBBARD, Rev. Mod. Phys. 33, 249 (1961).

39) T. HOLSTEIN, Annals of Physics 8, 343 (1959).

40) P. MORSE and H. FESHBACH, Methods of Theoretical Physics, McGraw-Hill, New York (1953).

41) B. FAIN, Phys. Stat. Sol. (b) 63, 411 (1974).

42) R.R. DOGONADZE, A.M. KUZNETSOV and M.A. VOROTYNSEV, Phys. Stat. Sol. (b) 54, 125 (1977).

43) B. FAIN, Chem. Phys. Lett. 56, 503 (1978).

44) S. LIN, J. Chem. Phys. 44, 3759 (1966).

45) K. FREED and J. JORTNER, J. Chem. Phys. 52, 6279 (1970).

46) R. ENGLMAN and J. JORTNER, Mol. Phys. 18, 145 (1980); K. FREED and J. JORTNER, J. Chem. Phys. 52, 6272 (1970); A. NITZAN and J. JORTNER, J. Chem. Phys. 58, 2412 (1973).

47) B. FAIN, Chem. Phys. Lett. 67, 267 (1979).

48) M. DIXON and E. WEBB, Enzymes, Longman, New York (1964); S.A. BERNHARD, The Structure and Function of Enzymes, New York (1968).

49) R.A. HARRIS and L. STODOLSKY, Phys. Lett. 78B, 313 (1978).

50) M. SIMONIUS, Phys. Rev. Lett. 40, 980 (1978).

51) D.S. TINTI and G.W. ROBINSON, J. Chem. Phys. 49, 3229 (1968); K. DRESSLER, O. OCHLER and D.A. SMITH, Phys. Rev. Lett. 34, 1364 (1975); H. DUBOST and R. CHARENEAU, Chem. Phys. 12, 407 (1976).

52) J. JORTNER, Phil. Mag. B, 40, 317 (1979).

53) V.M. KENKRE, Phys. Rev. A, 16, 766 (1977).

54) J. JORTNER, J. Chem. Phys. 64, 4860 (1976).

55) G.M. ZASLAVSKII and B.V. CHIRIKOV, Sov. Phys. Usp. 14, 549 (1972).

56) J. FORD, Adv. Chem. Phys. 24, 155 (1973).

57) D.W. OXTOBY and S. RICE, J. Chem. Phys. 65, 1676 (1976).

58) L. VAN HOVE, Physica 23, 441 (1957).

59) J.H. GIBBS, J. Chem. Phys. 57, 4473 (1972); J.H. GIBBS and P.D. FLEMING III, J. Stat. Phys. 12, 375 (1975); A.M. DUNKER, J.E. LUSK and J.H. GIBBS, Biophys. Chemistry 11, 9 (1980).

A. F. Williams

A Theoretical Approach to Inorganic Chemistry

1979. 144 figures, 17 tables. XII, 316 pages
ISBN 3-540-09073-8

This book is intended to outline the application of simple quantum mechanics to the study of inorganic chemistry, and to show its potential for systematizing and understanding the structure, physical properties, and reactivities of inorganic compounds. The considerable development of inorganic chemistry in recent years necessitates the establishment of a theoretical framework if the student is to acquire sound knowledge of the subject. An effort has been made to cover a wide range of subjects, and to encourage the reader to think of further extensions of the theories discussed. The importance of the critical application of theory is emphasized, and, although the book is concerned chiefly with molecular orbital theory, other approaches are discussed. The book is intended for students in the latter half of their undergraduate studies.

Contents: Quantum Mechanics and Atomic Theory. – Simple Molecular Orbital Theory. – Structural Applications of Molecular Orbital Theory. – Electronic Spectra and Magnetic Properties of Inorganic Compounds. – Alternative Methods and Concepts. – Mechanism and Reactivity. – Descriptive Chemistry. – Physical and Spectroscopic Methods. – Appendices. – Subject Index.

Springer-Verlag
Berlin
Heidelberg
New York